FUNDAMENTALS OF ALGEBRA

Edited by
John Kristtofer H. San Pedro

a
ArclerPress

Fundamentals of Algebra

Edited by John Kristtofer H. San Pedro

Edition 2018

ArclerPress

Published by

Arcler Press LLC

708 3rd Avenue, 6th Floor

New York

NY 10017

United States of America

www.arclerpress.com

ISBN: 978-1-68094-581-2 (Hardback)

Printed in United States of America

Library of Congress Control Number: 2017933146

About the Editor

John Kristtofer H. San Pedro

John is a Mechanical Engineer and a Licensed Master Plumber from the Polytechnic University of the Philippines. With a successful history in creating innovative, cost efficient designs. A career focus on building engineering with experience in conceptual design through production. Highly skilled in drafting and using design software's such as AutoCAD, Revit, and Solidworks for his works. Been actively involved in Kaizen programs and achieve to be a champion in such competition during his 2 years stay in the automotive/manufacturing field.

Table of Contents

3 NUMBER THEORY

4 GEOMETRY CONCEPTS

5 MATRIX

6 VECTOR SPACE

7 GEOMETRIC ALGEBRA

8 INNER PRODUCT SPACE

9 CATEGORY THEORY

10 LIST OF ABSTRACT ALGEBRA TOPICS

11 ALGEBRAIC STRUCTURES

List of Abbreviations

CGA	Conformal Geometric Algebra
GA	Geometric Algebra

Preface

Algebra is a branch of mathematics dealing with symbols and the rules for manipulating those symbols. In elementary algebra, those symbols represent quantities without fixed values, known as variables. Math cannot be fully understood or learned well without practicing. Algebra is no different. If you just remember all the rules and procedures without truly understanding the concepts, you will no doubt have difficulty learning algebra. Algebra can begin to make sense with the determination to take small steps, practice and increase the difficulty in the types of equations you solve in a step-by-step sequential process. Sometimes it helps to treat algebra like driving a car. You learn to drive with practice and there are certain rules to follow, knowing the rules makes driving easier and you avoid making mistakes. Algebra requires you to apply certain rules and the better you become at following the rules - which takes practice, the better you become a solving algebraic equations. Too often students think of algebra as an abstract topic completely disconnected from the real world. This may in part be attributed to the way in which many algebra curricula are written or presented, causing students to see the subject as valueless. Fortunately, real-life applications of algebra objectives abound, and learners can investigate them throughout the course. Interpreting and solving problems involving ratios, proportions and percents comprises another objective of algebra. Real-world scenarios in this genre can easily be created around the idea of store sales. Types of problems could include determining the percent off, percent saved, new cost or original cost. There are many additional real-life applications of algebra objectives. Exponential growth and decay functions connect to a wide variety of topics such as interest rates, population studies and bacteriology and medicine. Algebra objectives related to proportions may be used by cooks baking loaves of bread or dieters counting calories. This book explains how to do elementary algebra problems, solve equations and solve verbal problems.

Content Coverage

Chapters One and *Two* present an introduction to algebra and elementary algebra, respectively. Elementary algebra encompasses some of the basic concepts of algebra, one of the main branches of mathematics.

Chapters Three and *Four* are number theory and geometry basics, separately. Numer theory is devoted primarily to the study of the integers, whereas Geometry is a practical way for dealing with lengths, areas, and volumes.

Chapters Five and *Six* focus on matrix theory and vector space, correspondingly. Applications of matrices are found in most scientific fields. Vector spaces are the subject of linear algebra and are well characterized by their dimension, which, roughly speaking, specifies the number of independent directions in the space.

Chapters Seven and *Eight* clarifies about the concept of geometric algebra and inner product space, separately. A geometric algebra is a Clifford algebra of a vector space over the field of real numbers endowed with a quadratic form.

Chapter Nine emphasizes on category theory formalizes mathematical structure and its concepts in terms of a collection of objects and of arrows (also called morphisms).

Chapters Ten and *Eleven* present the fundamentals of abstract algebra and algebraic structures, respectively.

CHAPTER 1

Introduction to Algebra

Chapter Contents

Algebra is a branch of mathematics dealing with symbols and the rules for manipulating those symbols. In elementary algebra, those symbols (today written as Latin and Greek letters) represent quantities without fixed values, known as variables. Just as sentences describe relationships between specific words, in algebra, equations describe relationships between variables.

INTRODUCTION

Algebra (from Arabic *"al-jabr"* meaning "reunion of broken parts") is one of the broad parts of mathematics, together with number theory, geometry and analysis. In its most general form, algebra is the study of mathematical symbols and the rules for manipulating these symbols; it is a unifying thread of almost all of mathematics. As such, it includes everything from elementary equation solving to the study of abstractions such as groups, rings, and fields. The more basic parts of algebra are called elementary algebra, the more abstract parts are called abstract algebra or modern algebra. Elementary algebra is generally considered to be essential for any study of mathematics, science, or engineering, as well as such applications as medicine and economics. Abstract algebra is a major area in advanced mathematics, studied primarily by professional mathematicians.

$$x = \frac{-b \pm \sqrt{b^2 - 4ac}}{2a}$$

The quadratic formula expresses the solution of the degree two equation $ax^2 + bx + c = 0,$ in terms of its coefficients a, b, c, where a is not zero.

Elementary algebra differs from arithmetic in the use of abstractions, such as using letters to stand for numbers that are either unknown or allowed to take on many values. For example, in x+2=5 the letter x is unknown, but the law of inverses can be used to discover its value: x=3. In $E = mc^2$, the letters E and m are variables, and the letter c is a constant, the speed of light in a vacuum. Algebra gives methods for solving equations and expressing formulas that are much easier (for those who know how to use them) than the older method of writing everything out in words.

The word *algebra* is also used in certain specialized ways. A special kind of mathematical object in abstract algebra is called an "algebra", and the word is used, for example, in the phrases linear algebra and algebraic topology.

A mathematician who does research in algebra is called an algebraist.

ETYMOLOGY

The word *algebra* comes from the Arabic *al-jabr* "restoration" from the title of the book *Ilm al-jabr wa'l-muābala* by al-Khwarizmi. The word entered the English language during the fifteenth century, from either Spanish, Italian, or Medieval Latin. It originally referred to the surgical procedure of setting broken or dislocated bones. The mathematical meaning was first recorded in the sixteenth century.

DIFFERENT MEANINGS OF "ALGEBRA"

The word "algebra" has several related meanings in mathematics, as a single word or with qualifiers.

- As a single word without an article, "algebra" names a broad part of mathematics.
- As a single word with an article or in plural, "an algebra" or "algebras" denotes a specific mathematical structure, whose precise definition depends on the author. Usually the structure has an addition, multiplication, and a scalar multiplication. When some authors use the term "algebra", they make a subset of the following additional assumptions: associative, commutative, unital, and/or finite-dimensional. In universal algebra, the word "algebra" refers to a generalization of the above concept, which allows for n-ary operations.
- With a qualifier, there is the same distinction:
- Without an article, it means a part of algebra, such as linear algebra, elementary algebra (the symbol-manipulation rules taught in elementary courses of

mathematics as part of primary and secondary education), or abstract algebra (the study of the algebraic structures for themselves).

- With an article, it means an instance of some abstract structure, like a Lie algebra, an associative algebra, or a vertex operator algebra.
- Sometimes both meanings exist for the same qualifier, as in the sentence: *Commutative algebra is the study of commutative rings, which are commutative algebras over the integers.*

ALGEBRA AS A BRANCH OF MATHEMATICS

Algebra began with computations similar to those of arithmetic, with letters standing for numbers. This allowed proofs of properties that are true no matter which numbers are involved. For example, in the quadratic equation

$$ax^2 + bx + c = 0,$$

a, b, c can be any numbers whatsoever (except that a cannot be 0), and the quadratic formula can be used to quickly and easily find the values of the unknown quantity x satisfy the equation. That is to say, to find all the solutions of the equation.

Historically, and in current teaching, the study of algebra starts with the solving of equations such as the quadratic equation above. Then more general questions, such as "does an equation have a solution?", "how many solutions does an equation have?", "what can be said about the nature of the solutions?" are considered. These questions lead to ideas of form, structure and symmetry. This development permitted algebra to be extended to consider non-numerical objects, such as vectors, matrices, and polynomials. The structural properties of these non-numerical objects were then abstracted to define algebraic structures such as groups, rings, and fields.

Before the 16th century, mathematics was divided into only two subfields, arithmetic and geometry. Even though some methods, which had been developed much earlier, may be considered nowadays as algebra, the emergence of algebra and, soon thereafter, of infinitesimal calculus as subfields of mathematics only dates from the 16th or 17th century. From the second half of 19th century on, many new fields of mathematics appeared, most of which made use of both arithmetic and geometry, and almost all of which used algebra.

Today, algebra has grown until it includes many branches of mathematics, as can be seen in the Mathematics Subject Classification where none of the first level areas (two digit entries) is called *algebra*. Today algebra includes section 08-General algebraic systems, 12-Field theory and polynomials, 13-Commutative algebra, 15-Linear and multilinear algebra; matrix theory, 16-Associative rings and algebras, 17-Nonassociative rings and algebras, 18-Category theory; homological algebra, 19-K-theory and 20-Group theory. Algebra is also used extensively in 11-Number theory and 14-Algebraic geometry.

HISTORY

As a branch of mathematics, algebra emerged at the end of the 16th century in Europe, with the work of François Viète. Algebra can essentially be considered as doing computations similar to those of arithmetic but with non-numerical mathematical objects. However, until the 19th century, algebra consisted essentially of the theory of equations. For example, the fundamental theorem of algebra belongs to the theory of equations and is not, nowadays, considered as belonging to algebra.

This article describes the history of the theory of equations, called here "algebra", from the origins to the emergence of algebra as a separate area of mathematics.

Stages of Algebra

Algebraic Expression

Algebra did not always make use of the symbolism that is now ubiquitous in mathematics; instead, it went through three distinct stages. The stages in the development of symbolic algebra are approximately as follows:

- Rhetorical algebra, in which equations are written in full sentences. For example, the rhetorical form of $x + 1 = 2$ is "The thing plus one equals two" or possibly "The thing plus 1 equals 2". Rhetorical algebra was first developed by the ancient Babylonians and remained dominant up to the 16th century.
- Syncopated algebra, in which some symbolism is used, but which does not contain all of the characteristics of symbolic algebra. For instance, there may be a restriction that subtraction may be used only once within one side of an equation, which is not the case with symbolic algebra. Syncopated algebraic expression first appeared in Diophantus' *Arithmetica* (3rd century AD), followed by Brahmagupta's *Brahma Sphuta Siddhanta* (7th century).
- Symbolic algebra, in which full symbolism is used. Early steps toward this can be seen in the work of several Islamic mathematicians such as Ibn al-Banna (13th-14th centuries) and al-Qalasadi (15th century), although fully symbolic algebra was developed by François Viète (16th century). Later, René Descartes (17th century) introduced the modern notation (for example, the use of *x*—see below) and showed that the problems occurring in geometry can be expressed and solved in terms of algebra (Cartesian geometry).

Equally important as the use or lack of symbolism in algebra was the degree of the equations that were addressed. Quadratic equations played an important role in early algebra; and throughout most of history, until the early modern period, all quadratic equations were classified as belonging to one of three categories.

- $x^2 + px = q$
- $x^2 = px + q$
- $x^2 + q = px$

where p and q are positive. This trichotomy comes about because quadratic equations of the form $x^2 + px + q = 0$, with p and q positive, have no positive roots.

In between the rhetorical and syncopated stages of symbolic algebra, a geometric constructive algebra was developed by classical Greek and Vedic Indian mathematicians in which algebraic equations were solved through geometry. For instance, an equation of the form $x^2 = A$ was solved by finding the side of a square of area A.

Conceptual Stages

In addition to the three stages of expressing algebraic ideas, there were four conceptual stages in the development of algebra that occurred alongside the changes in expression. These four stages were as follows:

- Geometric stage, where the concepts of algebra are largely geometric. This dates back to the Babylonians and continued with the Greeks, and was later revived by Omar Khayyám.
- Static equation-solving stage, where the objective is to find numbers satisfying certain relationships. The move away from geometric algebra dates back to Diophantus and Brahmagupta, but algebra didn't decisively move to the static equation-solving stage until Al-Khwarizmi's *Al-Jabr*.
- Dynamic function stage, where motion is an underlying idea. The idea of a function began emerging with Sharaf al-Dīn al-Tūsī, but algebra did not decisively move to the dynamic function stage until Gottfried Leibniz.
- Abstract stage, where mathematical structure plays a central role. Abstract algebra is largely a product of the 19th and 20th centuries.

Babylonian Algebra

The origins of algebra can be traced to the ancient Babylonians, who developed a positional number system that greatly aided them in solving their rhetorical algebraic equations. The Babylonians were not interested in exact solutions but approximations, and so they would commonly use linear interpolation to approximate intermediate values. One of the most famous tablets is the Plimpton 322 tablet, created around 1900–1600 BCE, which gives a table of Pythagorean triples and represents some of the most advanced mathematics prior to Greek mathematics.

Babylonian algebra was much more advanced than the Egyptian algebra of the time; whereas the Egyptians were mainly concerned with linear equations the Babylonians were more concerned with quadratic and cubic equations. The Babylonians had developed flexible algebraic operations with which they were able to add equals to equals and multiply both sides of an equation by like quantities so as to eliminate fractions and factors. They were familiar with many simple forms of factoring, three-term quadratic equations with positive roots, and many cubic equations although it is not known if they were able to reduce the general cubic equation.

The Plimpton 322 tablet.

Egyptian Algebra

Ancient Egyptian algebra dealt mainly with linear equations while the Babylonians found these equations too elementary and developed mathematics to a higher level than the Egyptians.

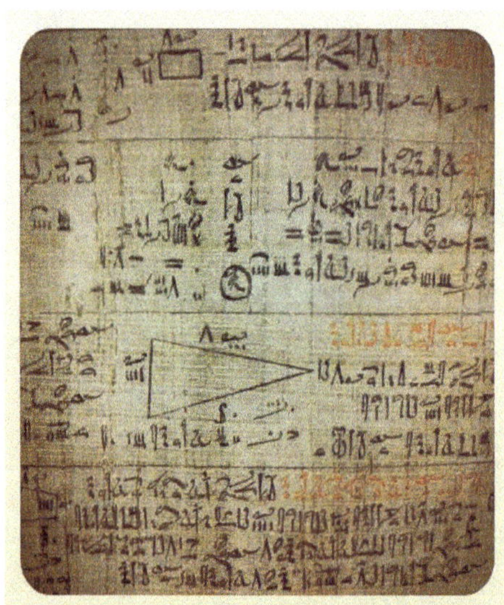

A portion of the Rhind Papyrus.

The Rhind Papyrus, also known as the Ahmes Papyrus, is an ancient Egyptian papyrus written c. 1650 BCE by Ahmes, who transcribed it from an earlier work that he dated to between 2000 and 1800 BCE. It is the most extensive ancient Egyptian mathematical document known to historians. The Rhind Papyrus contains problems where linear equations of the form $x + ax = b$ and $x + ax + bx = c$ are solved, where *a*, *b*, and *c* are known and *x*, which is referred to as "aha" or heap, is the unknown. The solutions were possibly, but not likely, arrived at by using the "method of false position", or *regula falsi*, where first a specific value is substituted into the left hand side of the equation, then the required arithmetic calculations are done, thirdly the result is compared to the right hand side of the equation, and finally the correct answer is found through the use of proportions. In some of the problems the author "checks" his solution, thereby writing one of the earliest known simple proofs.

Greek Geometric Algebra

It is sometimes alleged that the Greeks had no algebra, but this is inaccurate. By the time of Plato, Greek mathematics had undergone a drastic change. The Greeks created a geometric algebra where terms were represented by sides of geometric objects, usually lines, that had letters associated with them, and with this new form of algebra they were able to find solutions to equations by using a process that they invented, known as "the application of areas". "The application of areas" is only a part of geometric algebra and it is thoroughly covered in Euclid's *Elements*.

An example of geometric algebra would be solving the linear equation ax = bc. The ancient Greeks would solve this equation by looking at it as an equality of areas rather than as an equality between the ratios a:b and c:x. The Greeks would construct a rectangle with sides of length b and c, then extend a side of the rectangle to length a, and finally they would complete the extended rectangle so as to find the side of the rectangle that is the solution.

Bloom of Thymaridas

Iamblichus in *Introductio arithmatica* tells us that Thymaridas (c. 400 BCE – c. 350 BCE) worked with simultaneous linear equations. In particular, he created the then famous rule that was known as the "bloom of Thymaridas" or as the "flower of Thymaridas", which states that:

If the sum of *n* quantities be given, and also the sum of every pair containing a particular quantity, then this particular quantity is equal to 1/ (n - 2) of the difference between the sums of these pairs and the first given sum.

Ἐπὶ τῆς δοθείσης εὐθείας πεπερασμένης τρίγωνον ἰσόπλευρον συστήσασθαι.
Ἔστω ἡ δοθεῖσα εὐθεῖα πεπερασμένη ἡ ΑΒ.
Δεῖ δὴ ἐπὶ τῆς ΑΒ εὐθείας τρίγωνον ἰσόπλευρον συστήσασθαι.

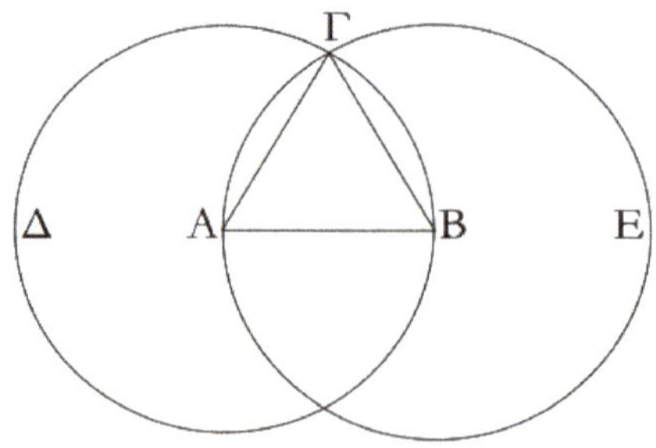

Κέντρῳ μὲν τῷ Α διαστήματι δὲ τῷ ΑΒ κύκλος γεγράφθω ὁ ΒΓΔ, καὶ πάλιν κέντρῳ μὲν τῷ Β διαστήματι δὲ τῷ ΒΑ κύκλος γεγράφθω ὁ ΑΓΕ, καὶ ἀπὸ τοῦ Γ σημείου, καθ᾽ ὃ τέμνουσιν ἀλλήλους οἱ κύκλοι, ἐπί τὰ Α, Β σημεῖα ἐπεζεύχθωσαν εὐθεῖαι αἱ ΓΑ, ΓΒ.

Καὶ ἐπεὶ τὸ Α σημεῖον κέντρον ἐστὶ τοῦ ΓΔΒ κύκλου, ἴση ἐστὶν ἡ ΑΓ τῇ ΑΒ· πάλιν, ἐπεὶ τὸ Β σημεῖον κέντρον ἐστὶ τοῦ ΓΑΕ κύκλου, ἴση ἐστὶν ἡ ΒΓ τῇ ΒΑ. ἐδείχθη δὲ καὶ ἡ ΓΑ τῇ ΑΒ ἴση· ἑκατέρα ἄρα τῶν ΓΑ, ΓΒ τῇ ΑΒ ἐστιν ἴση. τὰ δὲ τῷ αὐτῷ ἴσα καὶ ἀλλήλοις ἐστὶν ἴσα· καὶ ἡ ΓΑ ἄρα τῇ ΓΒ ἐστιν ἴση· αἱ τρεῖς ἄρα αἱ ΓΑ, ΑΒ, ΒΓ ἴσαι ἀλλήλαις εἰσίν.

Ἰσόπλευρον ἄρα ἐστὶ τὸ ΑΒΓ τρίγωνον. καὶ συνέσταται ἐπὶ τῆς δοθείσης εὐθείας πεπερασμένης τῆς ΑΒ. ὅπερ ἔδει ποιῆσαι.

A proof from Euclid's *Elements* that, given a line segment, an equilateral triangle exists that includes the segment as one of its sides.

or using modern notion, the solution of the following system of n linear equations in n unknowns,

$$x + x_1 + x_2 + \ldots + x_{n-1} = s$$

$$x + x_1 = m_1$$

$$x + x_2 = m_2$$

$$\cdot$$

$$\cdot$$

$$\cdot$$

$$x + x_{n-1} = m_{n-1}$$

is,

$$x = \frac{(m_1 + m_2 + \ldots + m_{n-1}) - s}{n-2} = \frac{(\sum_{i=1}^{n-1} m_i)}{n-2}$$

Iamblichus goes on to describe how some systems of linear equations that are not in this form can be placed into this form.

Euclid of Alexandria

Euclid was a Greek mathematician who flourished in Alexandria, Egypt, almost certainly during the reign of Ptolemy I (323–283 BCE). Neither the year nor place of his birth have been established, nor the circumstances of his death.

Hellenistic mathematician Euclid details geometrical algebra.

Euclid is regarded as the "father of geometry". His *Elements* is the most successful textbook in the history of mathematics.Although he is one of the most famous mathematicians in history there are no new discoveries attributed to him, rather he is remembered for his great explanatory skills. The *Elements* is not, as is sometimes thought, a collection of all Greek mathematical knowledge to its date, rather, it is an elementary introduction to it.

Elements

The geometric work of the Greeks, typified in Euclid's *Elements*, provided the framework for generalizing formulae beyond the solution of particular problems into more general systems of stating and solving equations.

Book II of the *Elements* contains fourteen propositions, which in Euclid's time were extremely significant for doing geometric algebra. These propositions and their results are the geometric equivalents of our modern symbolic algebra and trigonometry. Today, using modern symbolic algebra, we let symbols represent known and unknown magnitudes (i.e. numbers) and then apply algebraic operations on them. While in Euclid's time magnitudes were viewed as line segments and then results were deduced using the axioms or theorems of geometry.

Many basic laws of addition and multiplication are included or proved geometrically in the *Elements*. For instance, proposition 1 of Book II states:

If there be two straight lines, and one of them be cut into any number of segments whatever, the rectangle contained by the two straight lines is equal to the rectangles contained by the uncut straight line and each of the segments.

But this is nothing more than the geometric version of the (left) distributive law, $a(b+c+d)=ab+ac+ad$; and in Books V and VII of the *Elements* the commutative and associative laws for multiplication are demonstrated.

Many basic equations were also proved geometrically. For instance, proposition 5 in Book II proves that $a^2-b^2=(a+b)(a-b)$, and proposition 4 in Book II proves that $(a+b)^2=a^2+2ab+b^2$.

Furthermore, there are also geometric solutions given to many equations. For instance, proposition 6 of Book II gives the solution to the quadratic equation $ax + x^2 = b^2$, and proposition 11 of Book II gives a solution to $ax + x^2 = a^2$.

Data

Data is a work written by Euclid for use at the schools of Alexandria and it was meant to be used as a companion volume to the first six books of the *Elements*. The book contains some fifteen definitions and ninety-five statements, of which there are about two dozen statements that serve as algebraic rules or formulas. Some of these statements are geometric equivalents to solutions of quadratic equations. For instance, *Data* contains the solutions to the equations $dx^2 - adx + b^2c = 0$and the familiar Babylonian equation $xy = a^2$, $x \pm y = b$.

Conic Sections

A conic section is a curve that results from the intersection of a cone with a plane. There are three primary types of conic sections: ellipses (including circles), parabolas, and hyperbolas. The conic sections are reputed to have been discovered by Menaechmus (c. 380 BCE – c. 320 BCE) and since dealing with conic sections is equivalent to dealing with their respective equations, they played geometric roles equivalent to cubic equations and other higher order equations.

Menaechmus knew that in a parabola, the equation $y^2 = lx$ holds, where l is a constant called the latus rectum, although he was not aware of the fact that any equation in two unknowns determines a curve. He apparently derived these properties of conic sections and others as well. Using this information it was now possible to find a solution to the problem of the duplication of the cube by solving for the points at which two parabolas intersect, a solution equivalent to solving a cubic equation.

We are informed by Eutocius that the method he used to solve the cubic equation was due to Dionysodorus (250 BCE – 190 BCE). Dionysodorus solved the cubic by means of the intersection of a rectangular hyperbola and a parabola. This was related to a problem in Archimedes' *On the Sphere and Cylinder*. Conic sections would be studied and used for thousands of years by Greek, and later Islamic and European, mathematicians. In particular Apollonius of Perga's famous *Conics* deals with conic sections, among other topics.

Chinese Algebra

Chinese Mathematics dates to at least 300 BCE with the *Chou Pei Suan Ching*, generally considered to be one of the oldest Chinese mathematical documents.

Nine Chapters on the Mathematical Art

Chiu-chang suan-shu or *The Nine Chapters on the Mathematical Art*, written around 250 BCE, is one of the most influential of all Chinese math books and it is composed of some 246 problems.

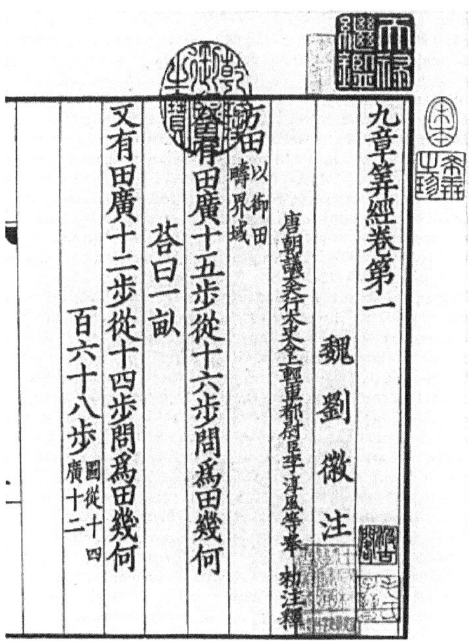

Nine Chapters on the Mathematical Art

Chapter eight deals with solving determinate and indeterminate simultaneous linear equations using positive and negative numbers, with one problem dealing with solving four equations in five unknowns.

Sea-Mirror of the Circle Measurements

Ts'e-yuan hai-ching, or *Sea-Mirror of the Circle Measurements*, is a collection of some 170 problems written by Li Zhi (or Li Ye) (1192 – 1279 CE). He used *fan fa*, or Horner's method, to solve equations of degree as high as six, although he did not describe his method of solving equations.

Mathematical Treatise in Nine Sections

Shu-shu chiu-chang, or *Mathematical Treatise in Nine Sections*, was written by the wealthy governor and minister Ch'in Chiu-shao (c. 1202 – c. 1261 CE) and with the invention of a method of solving simultaneous congruences, now called Chinese remainder theorem, it marks the high point in Chinese indeterminate analysis.

Magic Squares

The earliest known magic squares appeared in China. In *Nine Chapters* the author solves a system of simultaneous linear equations by placing the coefficients and constant terms of the linear equations into a magic square (i.e. a matrix) and performing column reducing operations on the magic square.

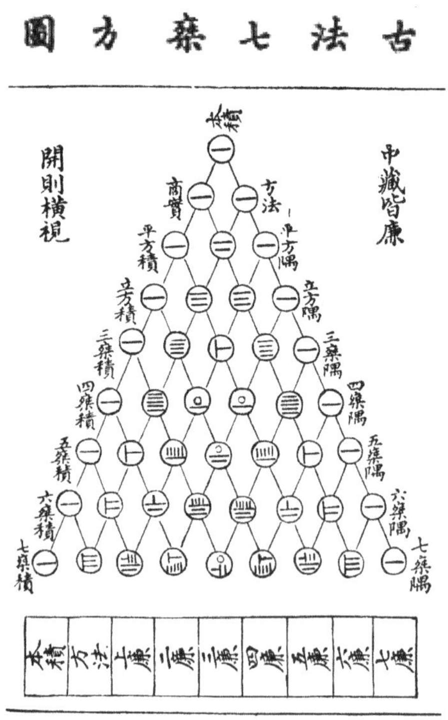

Yang Hui (Pascal's) triangle, as depicted by the ancient Chinese using rod numerals.

The earliest known magic squares of order greater than three are attributed to Yang Hui (fl. c. 1261 – 1275), who worked with magic squares of order as high as ten.

Precious Mirror of the Four Elements

Ssy-yüan yü-chien, or *Precious Mirror of the Four Elements*, was written by Chu Shih-chieh in 1303 and it marks the peak in the development of Chinese algebra. The four elements, called heaven, earth, man and matter, represented the four unknown quantities in his algebraic equations. The *Ssy-yüan yü-chien* deals with simultaneous equations and with equations of degrees as high as fourteen. The author uses the method of *fan fa*, today called Horner's method, to solve these equations.

The *Precious Mirror* opens with a diagram of the arithmetic triangle (Pascal's triangle) using a round zero symbol, but Chu Shih-chieh denies credit for it. A similar triangle appears in Yang Hui's work, but without the zero symbol.

There are many summation series equations given without proof in the *Precious mirror*. A few of the summation series are:

$$1^2 + 2^2 + 3^2 + \cdots + n^2 = \frac{n(n+1)(2n+1)}{3!}$$

$$1 + 8 + 30 + 80 + \cdots + \frac{n^2(n+1)(n+2)}{3!} = \frac{n(n+1)(n+2)(n+3)(4n+1)}{5!}$$

Diophantine Algebra

Diophantus was a Hellenistic mathematician who lived c. 250 CE, but the uncertainty of this date is so great that it may be off by more than a century. He is known for having written *Arithmetica*, a treatise that was originally thirteen books but of which only the first six have survived. *Arithmetica* has very little in common with traditional Greek mathematics since it is divorced from geometric methods, and it is different from Babylonian mathematics in that Diophantus is concerned primarily with exact solutions, both determinate and indeterminate, instead of simple approximations.

In *Arithmetica*, Diophantus is the first to use symbols for unknown numbers as well as abbreviations for powers of numbers, relationships, and operations; thus he used what is now known as *syncopated* algebra. The main difference between Diophantine syncopated algebra and modern algebraic notation is that the former lacked special symbols for operations, relations, and exponentials. So, for example, what we would write as

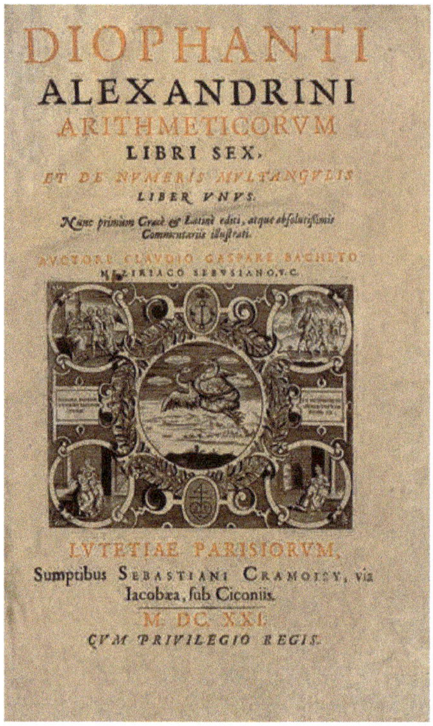

Cover of the 1621 edition of Diophantus' *Arithmetica*, translated into Latin by Claude Gaspard Bachet de Méziriac.

$$x^3 - 2x^2 + 10x - 1 = 5$$

Diophantus would have written this as
K^{Y} ας ι Δ^{Y} β **M** α σ **M** ε
where the symbols represent the following:

Symbol	Representation
α	represents 1
β	represents 2
ε	represents 5
ι	represents 10
ς	represents the unknown quantity (i.e. the variable)
σ	(short for σος) represents "equals"
	represents the subtraction of everything that follows it up to σ
M	represents the zeroth power of the variable (i.e. a constant term)
Δ^{Y}	represents the second power of the variable, from Greek δύναμις, meaning strength or power
K^{Y}	represents the third power of the variable, from Greek κύβος, meaning a cube

$\Delta^Y\Delta$	represents the fourth power of the variable
ΔK^Y	represents the fifth power of the variable
$K^Y K$	represents the sixth power of the variable

Note that the coefficients come after the variables and that addition is represented by the juxtaposition of terms. A literal symbol-for-symbol translation of Diophantus's syncopated equation into a modern symbolic equation would be the following:

$$x^3 1x10 - x^2 2x^0 1 = x^0 5$$

and, to clarify, if the modern parentheses and plus are used then the above equation can be rewritten as:

$$(x^3 1 + x10) - (x^2 2 + x^0 1) = x^0 5$$

Arithmetica is a collection of some 150 solved problems with specific numbers and there is no postulational development nor is a general method explicitly explained, although generality of method may have been intended and there is no attempt to find all of the solutions to the equations. *Arithmetica* does contain solved problems involving several unknown quantities, which are solved, if possible, by expressing the unknown quantities in terms of only one of them.*Arithmetica* also makes use of the identities:

$$\begin{aligned}(a^2 + b^2)(c^2 + d^2) &= (ac + db)^2 + (bc - ad)^2 \\ &= (ad + bc)^2 + (ac - bd)^2\end{aligned}$$

Indian Algebra

The Indian mathematicians were active in studying about number systems. The earliest known Indian mathematical documents are dated to around the middle of the first millennium BCE (around the 6th century BCE).

The recurring themes in Indian mathematics are, among others, determinate and indeterminate linear and quadratic equations, simple mensuration, and Pythagorean triples.

Aryabhata

Aryabhata (476–550 CE) was an Indian mathematician who authored *Aryabhatiya*. In it he gave the rules,

$$1^2 + 2^2 + \cdots + n^2 = \frac{n(n+1)(2n+1)}{6}$$

and

$$1^3 + 2^3 + \cdots + n^3 = (1 + 2 + \cdots + n)^2$$

Brahma Sphuta Siddhanta

Brahmagupta (fl. 628) was an Indian mathematician who authored *Brahma Sphuta Siddhanta*. In his work Brahmagupta solves the general quadratic equation for both positive and negative roots. In indeterminate analysis Brahmagupta gives the Pythagorean triads $m, \frac{1}{2}(\frac{m^2}{n} - n), \frac{1}{2}(\frac{m^2}{n} + n)$, but this is a modified form of an old Babylonian rule that Brahmagupta may have been familiar with. He was the first to give a general solution to the linear Diophantine equation ax + by = c, where a, b, and c are integers. Unlike Diophantus who only gave one solution to an indeterminate equation, Brahmagupta gave *all* integer solutions; but that Brahmagupta used some of the same examples as Diophantus has led some historians to consider the possibility of a Greek influence on Brahmagupta's work, or at least a common Babylonian source.

Like the algebra of Diophantus, the algebra of Brahmagupta was syncopated. Addition was indicated by placing the numbers side by side, subtraction by placing a dot over the subtrahend, and division by placing the divisor below the dividend, similar to our notation but without the bar. Multiplication, evolution, and unknown quantities were represented by abbreviations of appropriate terms. The extent of Greek influence on this syncopation, if any, is not known and it is possible that both Greek and Indian syncopation may be derived from a common Babylonian source.

Bhāskara II

Bhāskara II (1114 – c. 1185) was the leading mathematician of the 12th century. In Algebra, he gave the general solution of Pell's equation. He is the author of *Lilavati* and *Vija-Ganita*, which contain problems dealing with determinate and indeterminate linear and quadratic equations, and Pythagorean triples and he fails to distinguish between exact and approximate statements. Many of the problems in *Lilavati* and *Vija-Ganita* are derived from other Hindu sources, and so Bhaskara is at his best in dealing with indeterminate analysis.

Bhaskara uses the initial symbols of the names for colors as the symbols of unknown variables. So, for example, what we would write today as

$$(-x - 1) + (2x - 8) = x - 9$$

Bhaskara would have written as

ya $\dot{1}$ *ru* $\dot{1}$

ya 2 *ru* $\dot{8}$

Sum *ya* 1 ru $\dot{9}$

where *ya* indicates the first syllable of the word for *black*, and *ru* is taken from the word *species*. The dots over the numbers indicate subtraction.

Islamic Algebra

The first century of the Islamic Arab Empire saw almost no scientific or mathematical achievements since the Arabs, with their newly conquered empire, had not yet gained any intellectual drive and research in other parts of the world had faded. In the second half of the 8th century, Islam had a cultural awakening, and research in mathematics and the sciences increased. The Muslim Abbasid caliph al-Mamun (809–833) is said to have had a dream where Aristotle appeared to him, and as a consequence al-Mamun ordered that Arabic translation be made of as many Greek works as possible, including Ptolemy's *Almagest* and Euclid's *Elements*. Greek works would be given to the Muslims by the Byzantine Empire in exchange for treaties, as the two empires held an uneasy peace. Many of these Greek works were translated by Thabit ibn Qurra (826–901), who translated books written by Euclid, Archimedes, Apollonius, Ptolemy, and Eutocius.

There are three theories about the origins of Arabic Algebra. The first emphasizes Hindu influence, the second emphasizes Mesopotamian or Persian-Syriac influence and the third emphasizes Greek influence. Many scholars believe that it is the result of a combination of all three sources.

Throughout their time in power, before the fall of Islamic civilization, the Arabs used a fully rhetorical algebra, where often even the numbers were spelled out in words. The Arabs would eventually replace spelled out numbers (e.g. twenty-two) with Arabic numerals (e.g. 22), but the Arabs did not adopt or develop a syncopated or symbolic algebra until the work of Ibn al-Banna in the 13th century and Abū al-Hasan ibn Alī al-Qalasādī in the 15th century.

Al-jabr wa'l muqabalah

The Muslim Persian mathematician Muhammad ibn Mūsā al-Khwārizmī was a faculty member of the "House of Wisdom" (*Bait al-Hikma*) in Baghdad, which was established by Al-Mamun.

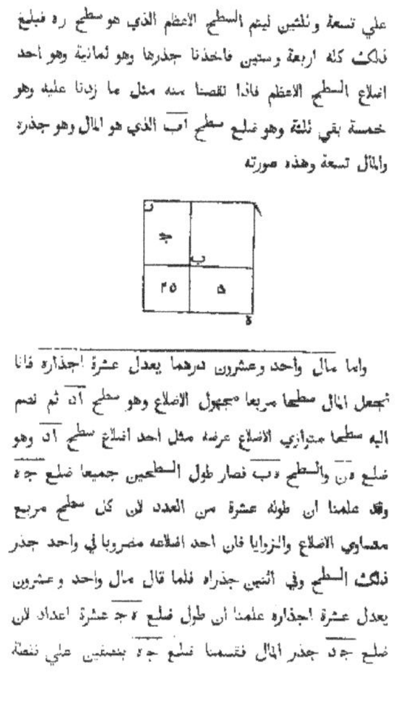

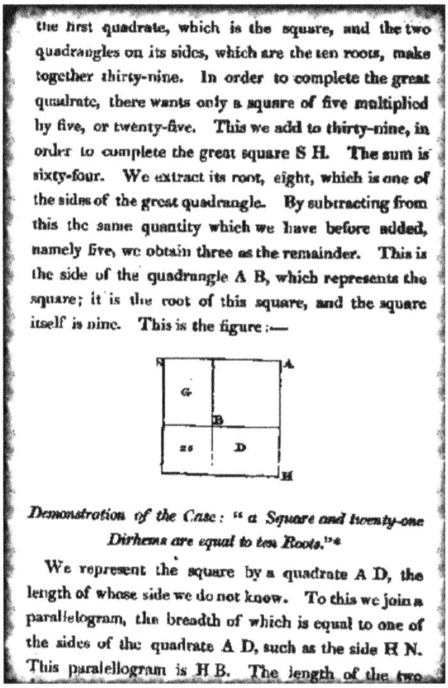
the first quadrate, which is the square, and the two quadrangles on its sides, which are the ten roots, make together thirty-nine. In order to complete the great quadrate, there wants only a square of five multiplied by five, or twenty-five. This we add to thirty-nine, in order to complete the great square S H. The sum is sixty-four. We extract its root, eight, which is one of the sides of the great quadrangle. By subtracting from this the same quantity which we have before added, namely five, we obtain three as the remainder. This is the side of the quadrangle A B, which represents the square; it is the root of this square, and the square itself is nine. This is the figure:—

Demonstration of the Case: "a Square and twenty-one Dirhems are equal to ten Roots."[4]

We represent the square by a quadrate A D, the length of whose side we do not know. To this we join a parallelogram, the breadth of which is equal to one of the sides of the quadrate A D, such as the side H N. This parallelogram is H B. The length of the two

Left: The original Arabic print manuscript of the Book of Algebra by Al-Khwarizmi. Right: A page from The Algebra of Al-Khwarizmi by Fredrick Rosen, in English.

Al-Khwarizmi, who died around 850 CE, wrote more than half a dozen mathematical and astronomical works, some of which were based on the Indian *Sindhind*. One of al-Khwarizmi's most famous books is entitled *Al-jabr wa'l muqabalah* or *The Compendious Book on Calculation by Completion and Balancing*, and it gives an exhaustive account of solving polynomials up to the second degree. The book also introduced the fundamental concept of "reduction" and "balancing", referring to the transposition of subtracted terms to the other side of an equation, that is, the cancellation of like terms on opposite sides of the equation. This is the operation which Al-Khwarizmi originally described as *al-jabr*.

R. Rashed and Angela Armstrong write:

"Al-Khwarizmi's text can be seen to be distinct not only from the Babylonian tablets, but also from Diophantus' *Arithmetica*. It no longer concerns a series of problems to be resolved, but an exposition which starts with primitive terms in which the combinations must give all possible prototypes for equations, which henceforward explicitly constitute the true object of study. On the other hand, the idea of an equation for its own sake appears from the beginning and, one could say, in a generic manner, insofar as it does not simply emerge in the course of solving a problem, but is specifically called on to define an infinite class of problems."

Al-Jabr is divided into six chapters, each of which deals with a different type of formula. The first chapter of *Al-Jabr* deals with equations whose squares equal its roots ($ax^2 = bx$), the second chapter deals with squares equal to number ($ax^2 = c$), the third chapter deals with roots equal to a number ($bx = c$), the fourth chapter deals with squares and roots equal a number ($ax^2 + bx = c$), the fifth chapter deals with squares and number equal roots ($ax^2 + c = bx$), and the sixth and final chapter deals with roots and number equal to squares ($bx + c = ax^2$).

Pages from a 14th-century Arabic copy of the book, showing geometric solutions to two quadratic equations

In *Al-Jabr*, al-Khwarizmi uses geometric proofs, he does not recognize the root $x = 0$, and he only deals with positive roots. He also recognizes that the discriminant must be positive and described the method of completing the square, though he does not justify the procedure. The Greek influence is shown by *Al-Jabr*'s geometric foundations and by one problem taken from Heron. He makes use of lettered diagrams but all of the coefficients in all of his equations are specific numbers since he had no way of expressing with parameters what he could express geometrically; although generality of method is intended.

Al-Khwarizmi most likely did not know of Diophantus's *Arithmetica*, which became known to the Arabs sometime before the 10th century. And even though al-Khwarizmi most likely knew of Brahmagupta's work, *Al-Jabr* is fully rhetorical with the numbers even being spelled out in words. So, for example, what we would write as

$$x^2 + 10x = 39$$

Diophantus would have written as

$\Delta^{Y}\alpha$ ςι ‘ίσ **M** λθ

And al-Khwarizmi would have written as

One square and ten roots of the same amount to thirty-nine *dirhems*; that is to say, what must be the square which, when increased by ten of its own roots, amounts to thirty-nine?

Logical Necessities in Mixed Equations

'Abd al-Hamīd ibn Turk authored a manuscript entitled *Logical Necessities in Mixed Equations*, which is very similar to al-Khwarzimi's *Al-Jabr* and was published at around the same time as, or even possibly earlier than, *Al-Jabr*. The manuscript gives exactly the same geometric demonstration as is found in *Al-Jabr*, and in one case the same example as found in *Al-Jabr*, and even goes beyond *Al-Jabr* by giving a geometric proof that if the discriminant is negative then the quadratic equation has no solution. The similarity between these two works has led some historians to conclude that Arabic algebra may have been well developed by the time of al-Khwarizmi and 'Abd al-Hamid.

Abu Kamil and al-Karkhi

Arabic mathematicians treated irrational numbers as algebraic objects. The Egyptian mathematician Abū Kāmil Shujā ibn Aslam (c. 850–930) was the first to accept irrational numbers (often in the form of a square root, cube root or fourth root) as solutions to quadratic equations or as coefficients in an equation. He was also the first to solve three non-linear simultaneous equations with three unknown variables.

Al-Karkhi (953–1029), also known as Al-Karaji, was the successor of Abū al-Wafā' al-Būzjānī (940–998) and he discovered the first numerical solution to equations of the form $ax^{2n} + bx^n = c$. Al-Karkhi only considered positive roots. Al-Karkhi is also regarded as the first person to free algebra from geometrical operations and replace them with the type of arithmetic operations which are at the core of algebra today. His work on algebra and polynomials, gave the rules for arithmetic operations to manipulate polynomials. The historian of mathematics F. Woepcke, in *Extrait du Fakhri, traité d'Algèbre par Abou Bekr Mohammed Ben Alhacan Alkarkhi* (Paris, 1853), praised Al-Karaji for being "the first who introduced the theory of algebraic calculus". Stemming from this, Al-Karaji investigated binomial coefficients and Pascal's triangle.

Omar Khayyám, Sharaf al-Dīn, and al-Kashi

Omar Khayyám (c. 1050 – 1123) wrote a book on Algebra that went beyond *Al-Jabr* to include equations of the third degree. Omar Khayyám provided both arithmetic and geometric solutions for quadratic equations, but he only gave geometric solutions for general cubic equations since he mistakenly believed that arithmetic solutions were impossible.His method of solving cubic equations by using intersecting conics had been used by Menaechmus, Archimedes, and Ibn al-Haytham (Alhazen), but Omar Khayyám generalized the method to cover all cubic equations with positive roots. He only considered positive roots and he did not go past the third degree. He also saw a strong relationship between Geometry and Algebra.

Omar Khayyam is credited with identifying the foundations of algebraic geometry and found the general geometric solution of the cubic equation.

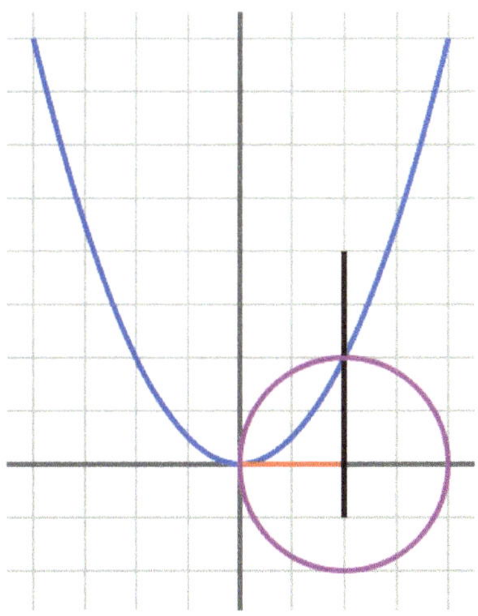

To solve the third-degree equation $x^3 + a^2x = b$ Khayyám constructed the parabola $x^2 = ay$, a circle with diameter b/a^2, and a vertical line through the intersection point. The solution is given by the length of the horizontal line segment from the origin to the intersection of the vertical line and the x-axis.

In the 12th century, Sharaf al-Dīn al-Tūsī (1135–1213) wrote the *Al-Mu'adalat* (*Treatise on Equations*), which dealt with eight types of cubic equations with positive solutions and five types of cubic equations which may not have positive solutions. He used what would later be known as the "Ruffini-Horner method" to numerically approximate the root of a cubic equation. He also developed the concepts of the maxima and minima of curves in order to solve cubic equations which may not have positive solutions. He understood the importance of the discriminant of the cubic equation and used an early version of Cardano's formula to find algebraic solutions to certain types of cubic equations. Some scholars, such as Roshdi Rashed, argue that Sharaf al-Din discovered the derivative of cubic polynomials and realized its significance, while other scholars connect his solution to the ideas of Euclid and Archimedes.

Sharaf al-Din also developed the concept of a function. In his analysis of the equation $x^3 + d = bx^2$ for example, he begins by changing the equation's form to $x^2(b - x) = d$. He then states that the question of whether the equation has a solution depends on whether or not the "function" on the left side reaches the value d. To determine this, he finds a maximum value for the function. He proves that the maximum value occurs when $x = \frac{2b}{3}$, which gives the functional value $\frac{4b^3}{27}$. Sharaf al-Din then states that if this value is less than d, there are no positive solutions; if it is equal to d then there is one solution at $\frac{2b}{3}$; and if it is greater than d, then there are two solutions, one between 0 and $\frac{2b}{3}$ and one between $\frac{2b}{3}$ and b.

In the early 15th century, Jamshīd al-Kāshī developed an early form of Newton's method to numerically solve the equation $x^P - N = 0$ to find roots of N. Al-Kāshī also developed decimal fractions and claimed to have discovered it himself. However, J. Lennart Berggrenn notes that he was mistaken, as decimal fractions were first used five centuries before him by the Baghdadi mathematician Abu'l-Hasan al-Uqlidisi as early as the 10th century.

Al-Hassār, Ibn al-Banna, and al-Qalasadi

Al-Hassār, a mathematician from Morocco specializing in Islamic inheritance jurisprudence during the 12th century, developed the modern symbolic mathematical notation for fractions, where the numerator and denominator are separated by a horizontal bar. This same fractional notation appeared soon after in the work of Fibonacci in the 13th century.

Abū al-Hasan ibn Alī al-Qalasādī (1412–1486) was the last major medieval Arab algebraist, who made the first attempt at creating an algebraic notation since Ibn al-Banna two centuries earlier, who was himself the first to make such an attempt since Diophantus and Brahmagupta in ancient times. The syncopated notations of his predecessors, however, lacked symbols for mathematical operations. Al-Qalasadi "took the first steps toward the introduction of algebraic symbolism by using letters in place of numbers" and by "using short Arabic words, or just their initial letters, as mathematical symbols."

European Algebra

Dark Ages

Just as the death of Hypatia signals the close of the Library of Alexandria as a mathematical center, so does the death of Boethius signal the end of mathematics in the Western Roman Empire. Although there was some work being done at Athens, it came to a close when in 529 the Byzantine emperor Justinian closed the pagan philosophical schools. The year 529 is now taken to be the beginning of the medieval period. Scholars fled the West towards the more hospitable East, particularly towards Persia, where they found haven under King Chosroes and established what might be termed an "Athenian Academy in Exile". Under a treaty with Justinian, Chosroes would eventually return the scholars to the Eastern Empire. During the Dark Ages, European mathematics was at its nadir with mathematical research consisting mainly of commentaries on ancient treatises; and most of this research was centered in the Byzantine Empire. The end of the medieval period is set as the fall of Constantinople to the Turks in 1453.

Late Middle Ages

The 12th century saw a flood of translations from Arabic into Latin and by the 13th century, European mathematics was beginning to rival the mathematics of other lands. In the 13th century, the solution of a cubic equation by Fibonacci is representative of the beginning of a revival in European algebra.

As the Islamic world was declining after the 15th century, the European world was ascending. And it is here that Algebra was further developed.

Modern Algebra

Modern notation for arithmetic operations was introduced between the end of the 15th century and the beginning of the 16th century by Johannes Widmann and Michael Stifel.

Another key event in the further development of algebra was the general algebraic solution of the cubic and quartic equations, developed in the mid-16th century. The idea of a determinant was developed by Japanese mathematician Kowa Seki in the 17th century, followed by Gottfried Leibniz ten years later, for the purpose of solving systems of simultaneous linear equations using matrices. Gabriel Cramer also did some work on matrices and determinants in the 18th century.

The symbol x

By tradition, the first unknown variable in an algebraic problem is nowadays represented by the symbol x; if there is a second or a third unknown, these are labeled y and z respectively. Algebraic x is conventionally printed in italic type to distinguish it from the sign of multiplication.

Mathematical historians generally agree that the use of x in algebra was introduced by René Descartes and was first published in his treatise *La Géométrie* (1637). In that work, he used letters from the beginning of the alphabet (a, b, c,...) for known quantities, and letters from the end of the alphabet (z, y, x,...) for unknowns. It has been suggested that he later settled on x (in place of z) for the first unknown because of its relatively greater abundance in the French and Latin typographical fonts of the time.

Three alternative theories of the origin of algebraic x were suggested in the 19th century: (1) a symbol used by German algebraists and thought to be derived from a cursive letter r, mistaken for x; (2) the numeral *1* with oblique strikethrough; and (3) an Arabic/Spanish source (see below). But the Swiss-American historian of mathematics Florian Cajori examined these and found all three lacking in concrete evidence; Cajori credited Descartes as the originator, and described his x, y, and z as "free from tradition[,] and their choice purely arbitrary."

Nevertheless, the Hispano-Arabic hypothesis continues to have a presence in popular culture today. It is the claim that algebraic x is the abbreviation of a supposed loanword from Arabic in Old Spanish. The theory originated in 1884 with the German "orientalist" Paul de Lagarde, shortly after he published his edition of a 1505 Spanish/Arabic bilingual glossary in which Spanish *cosa* ("thing") was paired with its Arabic equivalent, شىء (*shay*), transcribed as *xei*. (The "sh" sound in Old Spanish was routinely spelled x.) Evidently Lagarde was aware that Arab mathematicians, in the "rhetorical" stage of algebra's development, often used that word to represent the unknown quantity. He surmised that "nothing could be more natural" (Nichts war also natürlicher...) than for the initial of the Arabic word—romanized as the Old Spanish x—to be adopted for use in algebra. A later reader reinterpreted Lagarde's conjecture as having "proven" the point. Lagarde was unaware that early Spanish mathematicians used, not a *transcription* of the Arabic word, but rather its *translation* in their own language, "cosa". There is no instance of *xei* or similar forms in several compiled historical vocabularies of Spanish.

Gottfried Leibniz

Although the mathematical notion of function was implicit in trigonometric and logarithmic tables, which existed in his day, Gottfried Leibniz was the first, in 1692 and 1694, to employ it explicitly, to denote any of several geometric concepts derived from a curve, such as abscissa, ordinate, tangent, chord, and the perpendicular.In the 18th century, "function" lost these geometrical associations.

Leibniz realized that the coefficients of a system of linear equations could be arranged into an array, now called a matrix, which can be manipulated to find the solution of the system, if any. This method was later called Gaussian elimination. Leibniz also discovered Boolean algebra and symbolic logic, also relevant to algebra.

Abstract Algebra

The ability to do algebra is a skill cultivated in mathematics education. As explained by Andrew Warwick, Cambridge University students in the early 19th century practiced "mixed mathematics", doing exercises based on physical variables such as space, time, and weight. Over time the association of variables with physical quantities faded away as mathematical technique grew. Eventually mathematics was concerned completely with abstract polynomials, complex numbers, hypercomplex numbers and other concepts. Application to physical situations was then called applied mathematics or mathematical physics, and the field of mathematics expanded to include abstract algebra. For instance, the issue of constructible numbers showed some mathematical limitations, and the field of Galois theory was developed.

The Father of Algebra

The Hellenistic mathematician Diophantus has traditionally been known as "the father of algebra" but debate now exists as to whether or not Al-Khwarizmi deserves this title instead. Those who support Diophantus point to the fact that the algebra found in *Al-Jabr* is more elementary than the algebra found in *Arithmetica* and that *Arithmetica* is syncopated while *Al-Jabr* is fully rhetorical.

Those who support Al-Khwarizmi point to the fact that he gave an exhaustive explanation for the algebraic solution of quadratic equations with positive roots, and was the first to teach algebra in an elementary form and for its own sake, whereas Diophantus was primarily concerned with the theory of numbers. Al-Khwarizmi also introduced the fundamental concept of "reduction" and "balancing" (which he originally used the term *al-jabr* to refer to), referring to the transposition of subtracted terms to the other side of an equation, that is, the cancellation of like terms on opposite sides of the equation. Other supporters of Al-Khwarizmi point to his algebra no longer being concerned "with a series of problems to be resolved, but an exposition which starts with primitive terms in which the combinations must give all possible prototypes for equations, which henceforward explicitly constitute the true object of study." They also point to his treatment of an equation for its own sake and "in a generic manner, insofar as it does not simply emerge in the course of solving a problem, but is specifically called on to define an infinite class of problems."

AREAS OF MATHEMATICS WITH THE WORD ALGEBRA IN THEIR NAME

Some areas of mathematics that fall under the classification abstract algebra have the word algebra in their name; linear algebra is one example. Others do not: group theory, ring theory, and field theory are examples. In this section, we list some areas of mathematics with the word "algebra" in the name.

- Elementary algebra, the part of algebra that is usually taught in elementary courses of mathematics.
- Abstract algebra, in which algebraic structures such as groups, rings and fields are axiomatically defined and investigated.
- Linear algebra, in which the specific properties of linear equations, vector spaces and matrices are studied.
- Commutative algebra, the study of commutative rings.

- Computer algebra, the implementation of algebraic methods as algorithms and computer programs.
- Homological algebra, the study of algebraic structures that are fundamental to study topological spaces.
- Universal algebra, in which properties common to all algebraic structures are studied.
- Algebraic number theory, in which the properties of numbers are studied from an algebraic point of view.
- Algebraic geometry, a branch of geometry, in its primitive form specifying curves and surfaces as solutions of polynomial equations.
- Algebraic combinatorics, in which algebraic methods are used to study combinatorial questions.
- Relational algebra: a set of finitary relations that is closed under certain operators.

Many mathematical structures are called algebras:

- Algebra over a field or more generally algebra over a ring. Many classes of algebras over a field or over a ring have a specific name:
    - Associative algebra
    - Non-associative algebra
    - Lie algebra
    - Hopf algebra
    - C*-algebra
    - Symmetric algebra
    - Exterior algebra
    - Tensor algebra
- In measure theory,
    - Sigma-algebra
    - Algebra over a set
- In category theory
    - F-algebra and F-coalgebra
    - T-algebra
- In logic,
- Relation algebra, a residuated Boolean algebra expanded with an involution called converse.
- Boolean algebra, a structure abstracting the computation with the truth values *false* and *true*. The structures also have the same name.
- Heyting algebra

Elementary Algebra

Elementary algebra is the most basic form of algebra. It is taught to students who are presumed to have no knowledge of mathematics beyond the basic principles of arithmetic. In arithmetic, only numbers and their arithmetical operations (such as +, −, ×, ÷) occur.

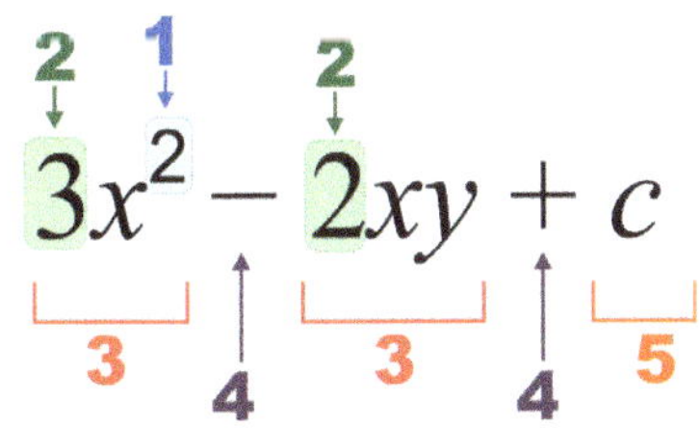

Algebraic expression notation:

1 – power (exponent)

2 – coefficient

3 – term

4 – operator

5 – constant term

x y c – variables/constants

In algebra, numbers are often represented by symbols called variables (such as *a*, *n*, *x*, *y* or *z*). This is useful because:

- It allows the general formulation of arithmetical laws (such as $a + b = b + a$ for all a and b), and thus is the first step to a systematic exploration of the properties of the real number system.
- It allows the reference to "unknown" numbers, the formulation of equations and the study of how to solve these. (For instance, "Find a number x such that $3x + 1 = 10$" or going a bit further "Find a number x such that $ax + b = c$". This step leads to the conclusion that it is not the nature of the specific numbers that allows us to solve it, but that of the operations involved.)
- It allows the formulation of functional relationships. (For instance, "If you sell x tickets, then your profit will be $3x - 10$ dollars, or $f(x) = 3x - 10$, where f is the function, and x is the number to which the function is applied".)

Polynomials

A polynomial is an expression that is the sum of a finite number of non-zero terms, each term consisting of the product of a constant and a finite number of variables raised to whole number powers. For example, $x^2 + 2x - 3$ is a polynomial in the single variable x. A polynomial expression is an expression that may be rewritten as a polynomial, by using commutativity, associativity and distributivity of addition and multiplication. For example, $(x - 1)(x + 3)$ is a polynomial expression, that, properly speaking, is not a polynomial. A polynomial function is a function that is defined by a polynomial, or, equivalently, by a polynomial expression. The two preceding examples define the same polynomial function.

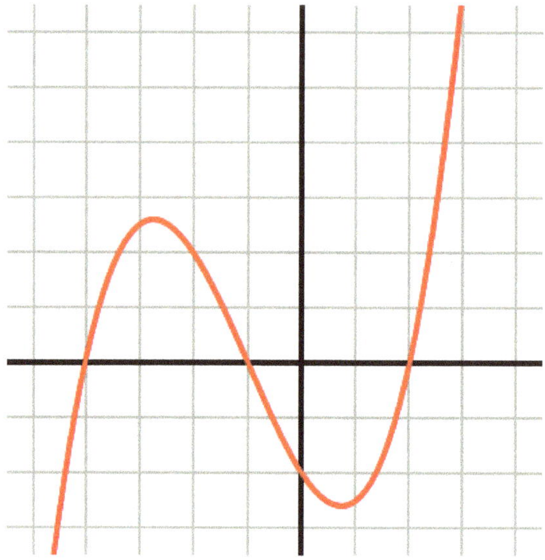

The graph of a polynomial function of degree 3.

Two important and related problems in algebra are the factorization of polynomials, that is, expressing a given polynomial as a product of other polynomials that can not be factored any further, and the computation of polynomial greatest common divisors. The example polynomial above can be factored as $(x - 1)(x + 3)$. A related class of problems is finding algebraic expressions for the roots of a polynomial in a single variable.

Education

It has been suggested that elementary algebra should be taught to students as young as eleven years old, though in recent years it is more common for public lessons to begin at the eighth grade level (≈ 13 y.o. ±) in the United States.

Since 1997, Virginia Tech and some other universities have begun using a personalized model of teaching algebra that combines instant feedback from specialized computer software with one-on-one and small group tutoring, which has reduced costs and increased student achievement.

Abstract Algebra

Abstract algebra extends the familiar concepts found in elementary algebra and arithmetic of numbers to more general concepts. Here are listed fundamental concepts in abstract algebra.

Sets: Rather than just considering the different types of numbers, abstract algebra deals with the more general concept of *sets*: a collection of all objects (called elements) selected by property specific for the set. All collections of the familiar types of numbers are sets. Other examples of sets include the set of all two-by-two matrices, the set of all second-degree polynomials ($ax^2 + bx + c$), the set of all two dimensional vectors in the plane, and the various finite groups such as the cyclic groups, which are the groups of integers modulo n. Set theory is a branch of logic and not technically a branch of algebra.

Binary operations: The notion of addition (+) is abstracted to give a *binary operation*, $*$ say. The notion of binary operation is meaningless without the set on which the operation is defined. For two elements a and b in a set S, $a * b$ is another element in the set; this condition is called closure. Addition (+), subtraction (−), multiplication (×), and division (÷) can be binary operations when defined on different sets, as are addition and multiplication of matrices, vectors, and polynomials.

Identity elements: The numbers zero and one are abstracted to give the notion of an *identity element* for an operation. Zero is the identity element for addition and one is the identity element for multiplication. For a general binary operator $*$ the identity element e must satisfy $a * e = a$ and $e * a = a$, and is necessarily unique, if it exists. This holds for addition as $a + 0 = a$ and $0 + a = a$ and multiplication $a \times 1 = a$ and $1 \times a = a$. Not all sets and operator combinations have an identity element; for example, the set of positive natural numbers (1, 2, 3, ...) has no identity element for addition.

Inverse elements: The negative numbers give rise to the concept of *inverse elements*. For addition, the inverse of a is written $-a$, and for multiplication the inverse is written a^{-1}. A general two-sided inverse element a^{-1} satisfies the property that $a * a^{-1} = e$ and $a^{-1} * a = e$, where e is the identity element.

Associativity: Addition of integers has a property called associativity. That is, the grouping of the numbers to be added does not affect the sum. For example: $(2 + 3) + 4 = 2 + (3 + 4)$. In general, this becomes $(a * b) * c = a * (b * c)$. This property is shared by most binary operations, but not subtraction or division or octonion multiplication.

Commutativity: Addition and multiplication of real numbers are both commutative. That is, the order of the numbers does not affect the result. For example: $2 + 3 = 3 + 2$. In general, this becomes $a * b = b * a$. This property does not hold for all binary operations. For example, matrix multiplication and quaternion multiplication are both non-commutative.

Groups

Combining the above concepts gives one of the most important structures in mathematics: a **group**. A group is a combination of a set S and a single binary operation $*$, defined in any way you choose, but with the following properties:

- An identity element e exists, such that for every member a of S, $e * a$ and $a * e$ are both identical to a.
- Every element has an inverse: for every member a of S, there exists a member a^{-1} such that $a * a^{-1}$ and $a^{-1} * a$ are both identical to the identity element.
- The operation is associative: if a, b and c are members of S, then $(a * b) * c$ is identical to $a * (b * c)$.

If a group is also commutative—that is, for any two members a and b of S, $a * b$ is identical to $b * a$—then the group is said to be abelian.

For example, the set of integers under the operation of addition is a group. In this group, the identity element is 0 and the inverse of any element a is its negation, $-a$. The associativity requirement is met, because for any integers a, b and c, $(a + b) + c = a + (b + c)$

The nonzero rational numbers form a group under multiplication. Here, the identity element is 1, since $1 \times a = a \times 1 = a$ for any rational number a. The inverse of a is $1/a$, since $a \times 1/a = 1$.

The integers under the multiplication operation, however, do not form a group. This is because, in general, the multiplicative inverse of an integer is not an integer. For example, 4 is an integer, but its multiplicative inverse is ¼, which is not an integer.

The theory of groups is studied in group theory. A major result in this theory is the classification of finite simple groups, mostly published between about 1955 and 1983, which separates the finite simple groups into roughly 30 basic types.

Semigroups, quasigroups, and monoids are structures similar to groups, but more general. They comprise a set and a closed binary operation, but do not necessarily satisfy the other conditions. A semigroup has an *associative* binary operation, but might not have an identity element. A monoid is a semigroup which does have an identity but might not have an inverse for every element. A quasigroup satisfies a requirement that any element can be turned into any other by either a unique left-multiplication or right-multiplication; however the binary operation might not be associative.

All groups are monoids, and all monoids are semigroups.

Examples										
Set	Natural numbers **N**		Integers **Z**		Rational numbers **Q** (also real **R** and complex **C** numbers)				Integers modulo 3: $\mathbf{Z}_3 = \{0, 1, 2\}$	
Operation	+	× (w/o zero)	+	× (w/o zero)	+	−	× (w/o zero)	÷ (w/o zero)	+	× (w/o zero)
Closed	Yes	Yes	Yes	Yes	Yes	Yes	Yes	Yes	Yes	Yes
Identity	0	1	0	1	0	N/A	1	N/A	0	1
Inverse	N/A	N/A	$-a$	N/A	$-a$	N/A	$1/a$	N/A	0, 2, 1, respectively	N/A, 1, 2, respectively
Associative	Yes	Yes	Yes	Yes	Yes	No	Yes	No	Yes	Yes
Commutative	Yes	Yes	Yes	Yes	Yes	No	Yes	No	Yes	Yes
Structure	monoid	monoid	abelian group	monoid	abelian group	quasigroup	abelian group	quasigroup	abelian group	abelian group ($\mathbf{Z}_2$)

Rings and Fields

Groups just have one binary operation. To fully explain the behavior of the different types of numbers, structures with two operators need to be studied. The most important of these are rings, and fields.

A ring has two binary operations (+) and (×), with × distributive over +. Under the first operator (+) it forms an *abelian group*. Under the second operator (×) it is associative, but it does not need to have identity, or inverse, so division is not required. The additive (+) identity element is written as 0 and the additive inverse of a is written as $-a$.

Distributivity generalizes the *distributive law* for numbers. For the integers $(a + b) \times c = a \times c + b \times c$ and $c \times (a + b) = c \times a + c \times b$, and × is said to be *distributive* over +.

The integers are an example of a ring. The integers have additional properties which make it an integral domain.

A field is a *ring* with the additional property that all the elements excluding 0 form an *abelian group* under ×. The multiplicative (×) identity is written as 1 and the multiplicative inverse of a is written as a^{-1}.

The rational numbers, the real numbers and the complex numbers are all examples of fields.

REFERENCES

1. "algebra". Online Etymology Dictionary.
2. I. N. Herstein, Topics in Algebra, "An algebraic system can be described as a set of objects together with some operations for combining them." p. 1, Ginn and Company, 1964
3. I. N. Herstein, Topics in Algebra, "...it also serves as the unifying thread which interlaces almost all of mathematics." p. 1, Ginn and Company, 1964
4. (Boyer 1991, "Europe in the Middle Ages" p. 258) "In the arithmetical theorems in Euclid's Elements VII-IX, numbers had been represented by line segments to which letters had been attached, and the geometric proofs in al-Khwarizmi's Algebra made use of lettered diagrams; but all coefficients in the equations used in the Algebra are specific numbers, whether represented by numerals or written out in words. The idea of generality is implied in al-Khwarizmi's exposition, but he had no scheme for expressing algebraically the general propositions that are so readily available in geometry."
5. "algebra". Oxford English Dictionary. Oxford University Press.
6. Gattengo, Caleb (2010). The Common Sense of Teaching Mathematics. Educational Solutions Inc. ISBN 978-0878252206.
7. "2010 Mathematics Subject Classification". Retrieved 5 October 2014.
8. Struik, Dirk J. (1987). A Concise History of Mathematics. New York: Dover Publications. ISBN 0-486-60255-9.
9. Boyer 1991
10. Cajori, Florian (2010). A History of Elementary Mathematics – With Hints on Methods of Teaching. p. 34. ISBN 1-4460-2221-8.
11. Roshdi Rashed (November 2009). "Al Khwarizmi: The Beginnings of Algebra". Saqi Books. ISBN 0-86356-430-5

12. "Diophantus, Father of Algebra". Retrieved 5 October 2014.
13. "History of Algebra". Retrieved 5 October 2014.
14. Josef W. Meri (2004). Medieval Islamic Civilization. Psychology Press. p. 31. ISBN 978-0-415-96690-0. Retrieved 25 November 2012.
15. Boyer, Carl B. (1991). A History of Mathematics (Second ed.). Wiley. pp. 178, 181. ISBN 0-471-54397-7.
16. Boyer, Carl B. (1991). A History of Mathematics (Second ed.). Wiley. p. 228. ISBN 0-471-54397-7.
17. (Boyer 1991, "The Arabic Hegemony" p. 229) "It is not certain just what the terms al-jabr and muqabalah mean, but the usual interpretation is similar to that implied in the translation above. The word al-jabr presumably meant something like "restoration" or "completion" and seems to refer to the transposition of subtracted terms to the other side of an equation; the word muqabalah is said to refer to "reduction" or "balancing" – that is, the cancellation of like terms on opposite sides of the equation."
18. (Boyer 1991, "The Arabic Hegemony" p. 230) "The six cases of equations given above exhaust all possibilities for linear and quadratic equations having positive root. So systematic and exhaustive was al-Khwarizmi's exposition that his readers must have had little difficulty in mastering the solutions."
19. Gandz and Saloman (1936), The sources of al-Khwarizmi's algebra, Osiris i, p. 263–277: "In a sense, Khwarizmi is more entitled to be called "the father of algebra" than Diophantus because Khwarizmi is the first to teach algebra in an elementary form and for its own sake, Diophantus is primarily concerned with the theory of numbers".
20. Rashed, R.; Armstrong, Angela (1994). The Development of Arabic Mathematics. Springer. pp. 11–2. ISBN 0-7923-2565-6. OCLC 29181926
21. Mathematical Masterpieces: Further Chronicles by the Explorers, p. 92
22. O'Connor, John J.; Robertson, Edmund F., "Sharaf al-Din al-Muzaffar al-Tusi", MacTutor History of Mathematics archive, University of St Andrews.
23. Victor J. Katz, Bill Barton; Barton, Bill (October 2007). "Stages in the History of Algebra with Implications for Teaching". Educational Studies in Mathematics. Springer Netherlands. 66 (2): 185–201 . doi:10.1007/s10649-006-9023-7
24. (Boyer 1991, "The Arabic Hegemony" p. 239) "Abu'l Wefa was a capable algebraist as well as a trigonometer. ... His successor al-Karkhi evidently used this translation to become an Arabic disciple of Diophantus – but without Diophantine analysis! ... In particular, to al-Karkhi is attributed the first numerical solution of equations of the form ax2n + bxn = c (only equations with positive roots were considered),"
25. "The Origins of Abstract Algebra". University of Hawaii Mathematics Department.
26. "The Collected Mathematical Papers".Cambridge University Press.
27. "Hull's Algebra" (pdf). New York Times. July 16, 1904. Retrieved September 21, 2012.

28. Quaid, Libby (September 22, 2008). "Kids misplaced in algebra" (Report). Associated Press. Retrieved September 23, 2012.
29. Hamilton, Reeve (7 September 2012). "THE TEXAS TRIBUNE; U.T.-Arlington Adopts New Way to Tackle Algebra". The New York Times. Retrieved 10 September 2012.
30. Boyer, Carl B. (1991), A History of Mathematics (Second ed.), John Wiley & Sons, Inc., ISBN 0-471-54397-7
31. Donald R. Hill, Islamic Science and Engineering (Edinburgh University Press, 1994).
32. Ziauddin Sardar, Jerry Ravetz, and Borin Van Loon, Introducing Mathematics (Totem Books, 1999).
33. George Gheverghese Joseph, The Crest of the Peacock: Non-European Roots of Mathematics (Penguin Books, 2000).
34. John J O'Connor and Edmund F Robertson, History Topics: Algebra Index. In MacTutor History of Mathematics archive (University of St Andrews, 2005).
35. I.N. Herstein: Topics in Algebra. ISBN 0-471-02371-X
36. R.B.J.T. Allenby: Rings, Fields and Groups. ISBN 0-340-54440-6
37. L. Euler: Elements of Algebra, ISBN 978-1-899618-73-6
38. Asimov, Isaac (1961). Realm of Algebra. Houghton Mifflin.

CHAPTER 2

Elementary Algebra

Chapter Contents

Elementary algebra, branch of mathematics that deals with the general properties of numbers and the relations between them. Algebra is fundamental not only to all further mathematics and statistics but to the natural sciences, computer science, economics, and business. Along with writing, it is a cornerstone of modern scientific and technological civilization. Earlier civilizations—Babylonian, Greek, Indian, Chinese, and Islamic—all contributed in important ways to the development of elementary algebra. It was left for Renaissance Europe, though, to develop an efficient system for representing all real numbers and a symbolism for representing unknowns, relations between them, and operations.

INTRODUCTION

Elementary algebra encompasses some of the basic concepts of algebra, one of the main branches of mathematics. It is typically taught to secondary school students and builds on their understanding of arithmetic. Whereas arithmetic deals with specified numbers, algebra introduces quantities without fixed values, known as variables. This use of variables entails a use of algebraic notation and an understanding of the general rules of the operators introduced in arithmetic. Unlike abstract algebra, elementary algebra is not concerned with algebraic structures outside the realm of real and complex numbers.

$$x = \frac{-b \pm \sqrt{b^2 - 4ac}}{2a}$$

The quadratic formula, which is the solution to the quadratic equation $ax^2 + bx + c = 0$, where $a \neq 0$. Here the symbols a, b, c represent arbitrary numbers, and x is a variable which represents the solution of the equation.

The use of variables to denote quantities allows general relationships between quantities to be formally and concisely expressed, and thus enables solving a broader scope of problems. Many quantitative relationships in science and mathematics are expressed as algebraic equations.

ALGEBRAIC NOTATION

Algebraic notation describes how algebra is written. It follows certain rules and conventions, and has its own terminology. For example, the expression $3x^2 - 2xy + c$ has the following components:

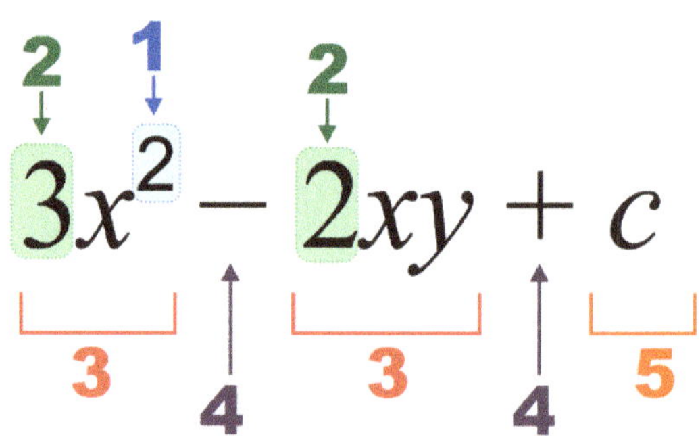

1 : Exponent (power), 2 : Coefficient, 3 : term, 4 : operator, 5 : constant, x, y: variables

A *coefficient* is a numerical value, or letter representing a numerical constant, that multiplies a variable (the operator is omitted). A *term* is an addend or a summand, a group of coefficients, variables, constants and exponents that may be separated from the other terms by the plus and minus operators. Letters represent variables and constants. By convention, letters at the beginning of the alphabet (e.g. a, b, c) are typically used to represent constants, and those toward the end of the alphabet (e.g. x, y and z) are used to represent variables. They are usually written in italics.

Algebraic operations work in the same way as arithmetic operations, such as addition, subtraction, multiplication, division and exponentiation. and are applied to algebraic variables and terms. Multiplication symbols are usually omitted, and implied when there is no space between two variables or terms, or when a coefficient is used. For example, $3 \times x^2$ is written as $3x^2$, and $2 \times x \times y$ may be written 2xy.

Usually terms with the highest power (exponent), are written on the left, for example, x^2 is written to the left of x. When a coefficient is one, it is usually omitted (e.g. $1x^2$ is written x^2). Likewise when the exponent (power) is one, (e.g. $3x^1$ is written 3x). When the exponent is zero, the result is always 1 (e.g. x^0 is always rewritten to 1). However 0^0, being undefined, should not appear in an expression, and care should be taken in simplifying expressions in which variables may appear in exponents.

Alternative Notation

Other types of notation are used in algebraic expressions when the required formatting is not available, or can not be implied, such as where only letters and symbols are available. For example, exponents are usually formatted using superscripts, e.g. x^2. In plain text, and in the TeX mark-up language, the caret symbol “” represents exponents, so x^2 is written as “x^2”. In programming languages such as Ada, Fortran, Perl, Python and Ruby, a double asterisk is used, so x^2 is written as “x**2”. Many programming languages and calculators use a single asterisk to represent the multiplication symbol, and it must be explicitly used, for example, 3x is written “3*x”.

CONCEPTS

Variables

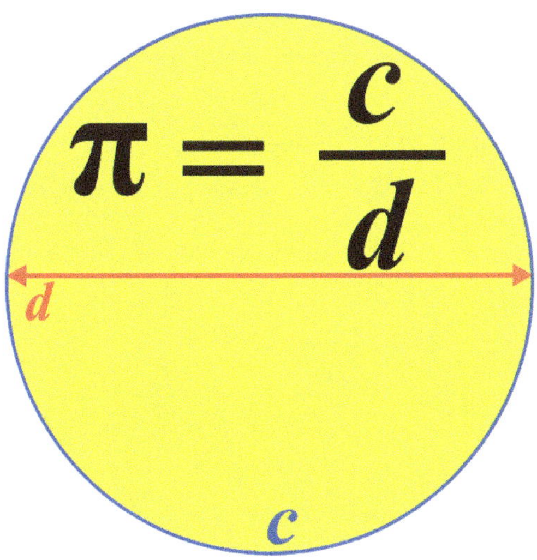

Example of variables showing the relationship between a circle’s diameter and its circumference. For any circle, its circumference c, divided by its diameter d is equal to the constant pi, π (approximately 3.14).

Elementary algebra builds on and extends arithmetic by introducing letters called variables to represent general (non-specified) numbers. This is useful for several reasons.

- Variables may represent numbers whose values are not yet known. For example, if the temperature of the current day, C, is 20 degrees higher than the temperature of the previous day, P, then the problem can be described algebraically as C=P+20.
- Variables allow one to describe *general* problems, without specifying the values of the quantities that are involved. For example, it can be stated specifically that 5 minutes is equivalent to $60 \times 5 = 300$seconds. A more general (algebraic) description may state that the number of seconds, $s = 60 \times m$, where m is the number of minutes.
- Variables allow one to describe mathematical relationships between quantities that may vary. For example, the relationship between the circumference, *c*, and diameter, *d*, of a circle is described by $\pi = c/d$..
- Variables allow one to describe some mathematical properties. For example, a basic property of addition is commutativity which states that the order of numbers being added together does not matter. Commutativity is stated algebraically as $(a + b) = (b + a)$.

Evaluating Expressions

Algebraic expressions may be evaluated and simplified, based on the basic properties of arithmetic operations (addition, subtraction, multiplication, division and exponentiation). For example,

- Added terms are simplified using coefficients. For example, x+x+x can be simplified as 3x (where 3 is a numerical coefficient).
- Multiplied terms are simplified using exponents. For example, x+x+x is represented as x^3
- Like terms are added together, for example, $2x^2 + 3ab - x^2 + ab$ is written as $x^2 + 4ab$, because the terms containing x^2 are added together, and, the terms containing ab are added together.
- Brackets can be "multiplied out", using the distributive property. For example, $x(2x + 3)$ can be written as $(x \times 2x) + (x \times 3)$ which can be written as $2x^2 + 3x$
- Expressions can be factored. For example, $6x^5 + 3x^2$, by dividing both terms $3x^2$ can be written as $3x^2(2x^3 + 1)$

Equations

An equation states that two expressions are equal using the symbol for equality, = (the equals sign). One of the most well-known equations describes Pythagoras' law relating the length of the sides of a right angle triangle:

$$c^2 = a^2 + b^2$$

This equation states that c^2, representing the square of the length of the side that is the hypotenuse (the side opposite the right angle), is equal to the sum (addition) of the squares of the other two sides whose lengths are represented by a and b.

An equation is the claim that two expressions have the same value and are equal. Some equations are true for all values of the involved variables (such as $a + b = b + a$); such equations are called identities. Conditional equations are true for only some values of the involved variables, e.g. $x^2 - 1 = 8$ is true only for $x = 3$ and $x = -3$. The values of the variables which make the equation true are the solutions of the equation and can be found through equation solving.

Another type of equation is an inequality. Inequalities are used to show that one side of the equation is greater, or less, than the other. The symbols used for this are: $a > b$ where $>$ represents 'greater than', and $a < b$ where $<$ represents 'less than'. Just like standard equality equations, numbers can be added, subtracted, multiplied or divided. The only exception is that when multiplying or dividing by a negative number, the inequality symbol must be flipped.

Properties of Equality

By definition, equality is an equivalence relation, meaning it has the properties (a) reflexive (i.e.b=b), (b) symmetric (i.e. if a=b then b=a) (c) transitive (i.e. if a=b and b=c then a=c). It also satisfies the important property that if two symbols are used for equal things, then one symbol can be substituted for the other in any true statement about the first and the statement will remain true. This implies the following properties:

- if $a = b$ and $c = d$ then $a + c = b + d$ and $ac = bd$;
- if $a = b$ then $a + c = b + c$;
- more generally, for any function f, if $a = b$ then $f(a) = f(b)$.

Properties of Inequality

The relations *less than* < and greater than > have the property of transitivity:

If $a < b$ and $b < c$ then $a < c$;

If $a < b$ and $c < d$ then $a + c < b + d$;

If $a < b$ and $c > 0$ then $ac < bc$;

If $a < b$ and $c < 0$ then $bc < ac$.

By reversing the inequation, < and > can be swapped, for example:

- $a < b$ is equivalent to $b > a$

Substitution

Substitution is replacing the terms in an expression to create a new expression. Substituting 3 for a in the expression a*5 makes a new expression 3*5 with meaning 15. Substituting the terms of a statement makes a new statement. When the original statement is true independent of the values of the terms, the statement created by substitutions is also true. Hence definitions can be made in symbolic terms and interpreted through substitution: if a informs the reader of this statement that 3^2 informs the reader of this statement that x means 3*3=9. Often it's not known whether the statement is true independent of the values of the terms, and substitution allows one to derive restrictions on the possible values, or show what conditions the statement holds under. For example, taking the statement x+1=0, if x is substituted with 1, this imples 1+1=2=0, which is false, which implies that if x+1=0 then x can't be 1.

If x and y are integers, rationals, or real numbers, then $xy=0$ implies $x=0$ or $y=0$. Suppose $abc=0$. Then, substituting a for x and bc for y, we learn $a=0$ or $bc=0$. Then we can substitute again, letting $x=b$ and $y=c$, to show that if $bc=0$ then $b=0$ or $c=0$. Therefore, if $abc=0$, then $a=0$ or ($b=0$ or $c=0$), so $abc=0$ implies $a=0$ or $b=0$ or $c=0$.

Consider if the original fact were stated as "$ab=0$ implies $a=0$ or $b=0$." Then when we say "suppose $abc=0$," we have a conflict of terms when we substitute. Yet the above logic is still valid to show that if $abc=0$ then $a=0$ or $b=0$ or $c=0$ if instead of letting $a=a$ and $b=bc$ we substitute a for a and b for bc (and with $bc=0$, substituting b for a and c for b). This shows that substituting for the terms in a statement isn't always the same as letting the terms from the statement equal the substituted terms. In this situation it's clear that if we substitute an expression a into the a term of the original equation, the a substituted does not refer to the a in the statement "$ab=0$ implies $a=0$ or $b=0$."

SOLVING ALGEBRAIC EQUATIONS

The following sections lay out examples of some of the types of algebraic equations that may be encountered.

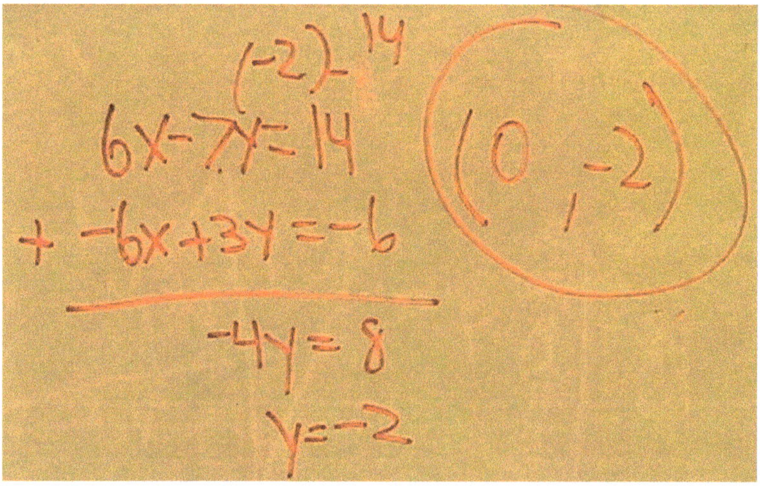

A typical algebra problem.

Linear Equations with One Variable

Linear equations are so-called, because when they are plotted, they describe a straight line. The simplest equations to solve are linear equations that have only one variable. They contain only constant numbers and a single variable without an exponent. As an example, consider:

Problem in words: If you double the age of a child and add 4, the resulting answer is 12. How old is the child?

Linear Equations with One Variable

Linear equations are so-called, because when they are plotted, they describe a straight line. The simplest equations to solve are linear equations that have only one variable. They contain only constant numbers and a single variable without an exponent. As an example, consider:

Problem in words: If you double the age of a child and add 4, the resulting answer is 12. How old is the child?

Equivalent equation: $2x + 4 = 12$ where x represent the child's age

To solve this kind of equation, the technique is add, subtract, multiply, or divide both sides of the equation by the same number in order to isolate the variable on one side of the equation. Once the variable is isolated, the other side of the equation is the value of the variable. This problem and its solution are as follows:

1. Equation to solve: $2x + 4 = 12$
2. Subtract 4 from both sides: $2x + 4 - 4 = 12 - 4$
3. This simplifies to: $2x = 8$
4. Divide both sides by 2: $\frac{2x}{2} = \frac{8}{2}$
5. This simplifies to the solution: $x = 4$

The general form of a linear equation with one variable, can be written as: $ax + b = c$

Following the same procedure (i.e. subtract b from both sides, and then divide by a), the general solution is given by

$$x = \frac{c - b}{a}$$

Linear Equations with Two Variables

A linear equation with two variables has many (i.e. an infinite number of) solutions.For example:

Problem in words: A father is 22 years older than his son. How old are they?

Equivalent equation: $y = x + 22$ where y is the father's age, x is the son's age.

This cannot be worked out by itself. If the son's age was made known, then there would no longer be two unknowns (variables), and the problem becomes a linear equation with just one variable, that can be solved as described above.

To solve a linear equation with two variables (unknowns), requires two related equations. For example, if it was also revealed that:

Problem in words:	In 10 years, the father will be twice as old as his son.
Equivalent equation:	$y + 10 = 2 \times (x + 10)$
Subtract 10 from both sides:	$y = 2 \times (x + 10) - 10$
Multiple out brackets:	$y = 2x + 20 - 10$
Simplify:	$y = 2x + 10$

Now there are two related linear equations, each with two unknowns, which enables the production of a linear equation with just one variable, by subtracting one from the other (called the elimination method):

Second equation	$y = 2x + 10$
First equation	$y = x + 22$
Subtract the first equation from the second in order to remove y	$(y - y) = (2x - x) + 10 - 22$
Simplify	$0 = x - 12$
Add 12 to both sides	$12 = x$
Rearrange	$x = 12$

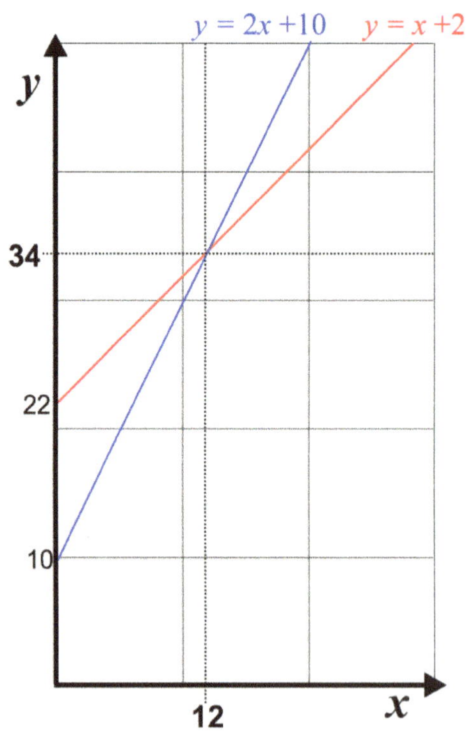

Solving two linear equations with a unique solution at the point that they intersect.

In other words, the son is aged 12, and since the father 22 years older, he must be 34. In 10 years time, the son will be 22, and the father will be twice his age, 44. This problem is illustrated on the associated plot of the equations.

For other ways to solve this kind of equations, see below, System of linear equations.

Quadratic Equations

A quadratic equation is one which includes a term with an exponent of 2, for example, x^2, and no term with higher exponent. The name derives from the Latin *quadrus*, meaning square. In general, a quadratic equation can be expressed in the form $ax^2 + bx + c = 0$, where a is not zero (if it were zero, then the equation would not be quadratic but linear). Because of this a quadratic equation must contain the term ax^2, which is known as the quadratic term. Hence $a \neq 0$, and so we may divide by a and rearrange the equation into the standard form

$$x^2 + px + q = 0$$

where $p = b/a$ and $q = c/a$. Solving this, by a process known as completing the square, leads to the quadratic formula

$$x = \frac{-b \pm \sqrt{b^2 - 4ac}}{2a},$$

where the symbol "±" indicates that both

$$x = \frac{-b + \sqrt{b^2 - 4ac}}{2a} \quad \text{and} \quad x = \frac{-b - \sqrt{b^2 - 4ac}}{2a}$$

are solutions of the quadratic equation.

Quadratic equations can also be solved using factorization (the reverse process of which is expansion, but for two linear terms is sometimes denoted foiling). As an example of factoring:

$$x^2 + 3x - 10 = 0,$$

which is the same thing as

$$(x + 5)(x - 2) = 0.$$

It follows from the zero-product property that either $x = 2$ or $x = -5$ are the solutions, since precisely one of the factors must be equal to zero. All quadratic equations will have two solutions in the complex number system, but need not have any in the real number system. For example,

$$x^2 + 1 = 0$$

has no real number solution since no real number squared equals −1. Sometimes a quadratic equation has a root of multiplicity 2, such as:

$$(x + 1)^2 = 0.$$

For this equation, −1 is a root of multiplicity 2. This means −1 appears two times, since the equation can be rewritten in factored form as

$$[x - (-1)][x - (-1)] = 0.$$

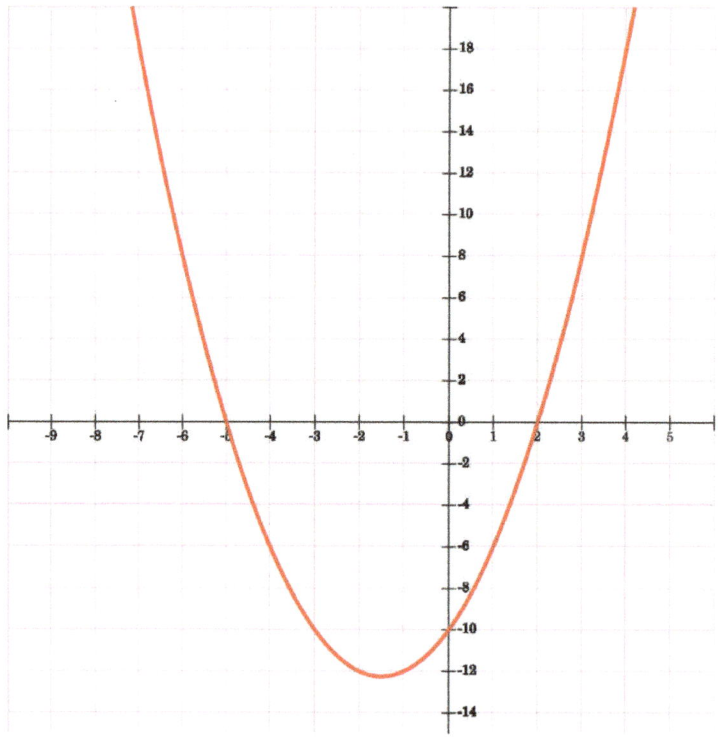

Quadratic equation plot of $y = x^2 + 3x - 10$ showing its roots at $x = -5$ and $x = 2$, and that the quadratic can be rewritten as $y = (x+5)(x-2)$

Complex Numbers

All quadratic equations have two solutions in complex numbers, a category that includes real numbers, imaginary numbers, and sums of real and imaginary numbers. Complex numbers first arise in the teaching of quadratic equations and the quadratic formula. For example, the quadratic equation

$$x^2 + x + 1 = 0$$

has solutions

$$x = \frac{-1 + \sqrt{-3}}{2} \quad \text{and} \quad x = \frac{-1 - \sqrt{-3}}{2}.$$

Since $\sqrt{-3}$ is not any real number, both of these solutions for x are complex numbers.

Exponential and Logarithmic Equations

An exponential equation is one which has the form $a^x = b$ for $a > 0$, which has solution

$$X = \log_a b = \frac{\ln b}{\ln a}$$

when $b > 0$. Elementary algebraic techniques are used to rewrite a given equation in the above way before arriving at the solution. For example, if

$$3 \cdot 2^{x-1} + 1 = 10$$

then, by subtracting 1 from both sides of the equation, and then dividing both sides by 3 we obtain

$$2^{x-1} = 3$$

whence

$$x - 1 = \log_2 3$$

or

$$x = \log_2 3 + 1.$$

A logarithmic equation is an equation of the form $log_a(x) = b$ for $a > 0$, which has solution

$$X = a^b.$$

For example, if

$$4\log_5(x - 3) - 2 = 6$$

then, by adding 2 to both sides of the equation, followed by dividing both sides by 4, we get

$$\log_5(x - 3) = 2$$

whence

$$x - 3 = 5^2 = 25$$

from which we obtain $x = 28.$

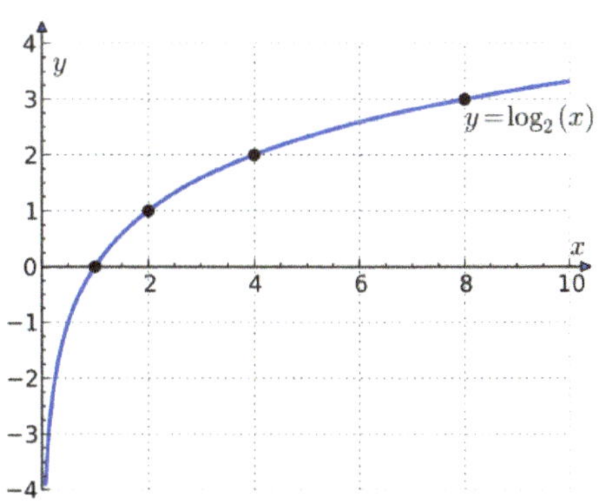

The graph of the logarithm to base 2 crosses the x axis (horizontal axis) at 1 and passes through the points with coordinates (2, 1), (4, 2), and (8, 3). For example, $\log_2(8) = 3$, because $2^3 = 8$. The graph gets arbitrarily close to the y axis, but does not meet or intersect it.

Radical Equations

A radical equation is one that includes a radical sign, which includes square roots, $\sqrt{x}$, cube roots, $\sqrt[3]{x}$, and nth roots, $\sqrt[n]{x}$. Recall that an nth root can be rewritten in exponential format, so that $\sqrt[n]{x}$ is equivalent to $x^{\frac{1}{n}}$. Combined with regular exponents (powers), then $\sqrt[2]{x^3}$ (the square root of {\displaystyle x} x cubed), can be rewritten as $x^{\frac{3}{2}}$ So a common form of a radical equation is $\sqrt[n]{x^m} = a$ (equivalent to $x^{\frac{m}{n}} = a$) where m and n are integers. It has real solution(s):

$$\sqrt[2]{x^3} \equiv x^{\frac{3}{2}}$$

Radical equation showing two ways to represent the same expression. The triple bar means the equation is true for all values of x

m is odd	m is even and $a \geq 0$	m and n are even and $a < 0$	m is even, n is odd, and $a < 0$
$x = \sqrt[m]{a^n}$ equivalently $x = \left(\sqrt[m]{a}\right)^n$	$x = \pm\sqrt[m]{a^n}$ equivalently $x = \pm\left(\sqrt[m]{a}\right)^n$	$x = \pm\sqrt[m]{a^n}$	no real solution

For example, if:

$$(x+5)^{2/3} = 4,$$

then

$$x + 5 = \pm(\sqrt{4})^3$$
$$x + 5 = \pm 8$$
$$x = -5 \pm 8$$
$$x = 3, -13$$

System of Linear Equations

There are different methods to solve a system of linear equations with two variables.

Elimination Method

An example of solving a system of linear equations is by using the elimination method:

$$\begin{cases} 4x + 2y &= 14 \\ 2x - y &= 1. \end{cases}$$

Multiplying the terms in the second equation by 2:

$$4x + 2y = 14$$
$$4x - 2y = 2.$$

Adding the two equations together to get:

$$8x = 16$$

which simplifies to

$$x = 2.$$

Since the fact that $x = 2$ is known, it is then possible to deduce that $y = 3$ by either of the original two equations (by using *2* instead of x The full solution to this problem is then

$$\begin{cases} x = 2 \\ y = 3. \end{cases}$$

Note that this is not the only way to solve this specific system; y could have been solved before x.

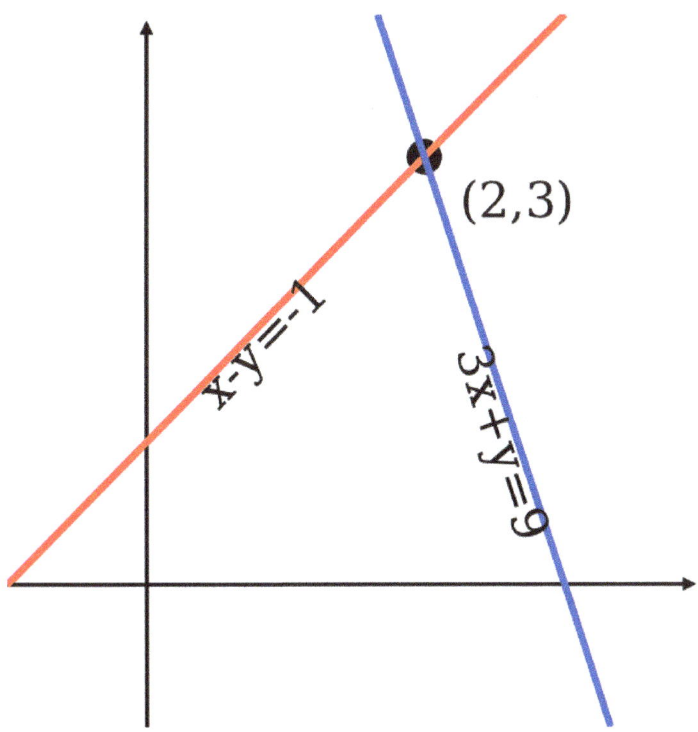

The solution set for the equations x - y = -1 and 3x + y = 9 is the single point (2, 3).

Substitution Method

Another way of solving the same system of linear equations is by substitution.

$$\begin{cases} 4x + 2y &= 14 \\ 2x - y &= 1. \end{cases}$$

An equivalent for y can be deduced by using one of the two equations. Using the second equation:

$$2x - y = 1$$

Subtracting $2x$ from each side of the equation:

$$\begin{aligned} 2x - 2x - y &= 1 - 2x \\ -y &= 1 - 2x \end{aligned}$$

and multiplying by −1:

$$y = 2x - 1.$$

Using this y value in the first equation in the original system:

$$\begin{aligned} 4x + 2(2x - 1) &= 14 \\ 4x + 4x - 2 &= 14 \\ 8x - 2 &= 14 \end{aligned}$$

Adding *2* on each side of the equation:

$$\begin{aligned} 8x - 2 + 2 &= 14 + 2 \\ 8x &= 16 \end{aligned}$$

which simplifies to

$$x = 2$$

Using this value in one of the equations, the same solution as in the previous method is obtained.

$$\begin{cases} x = 2 \\ y = 3. \end{cases}$$

Note that this is not the only way to solve this specific system; in this case as well, y could have been solved before x.

Other Types of Systems of Linear Equations

Inconsistent Systems

In the above example, a solution exists. However, there are also systems of equations which do not have any solution. Such a system is called inconsistent. An obvious example is

$$\begin{cases} x + y = 1 \\ 0x + 0y = 2\,. \end{cases}$$

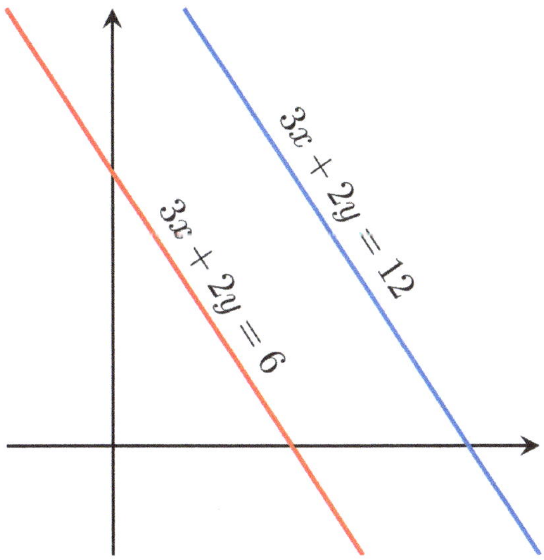

The equations $3x + 2y = 6$ and $3x + 2y = 12$ are parallel and cannot intersect, and is unsolvable.

As 0≠2, the second equation in the system has no solution. Therefore, the system has no solution. However, not all inconsistent systems are recognized at first sight. As an example, let us consider the system

$$\begin{cases} 4x + 2y = 12 \\ -2x - y = -4. \end{cases}$$

Multiplying by 2 both sides of the second equation, and adding it to the first one results in

$$0x + 0y = 4,$$

which has clearly no solution.

Undetermined Systems

There are also systems which have infinitely many solutions, in contrast to a system with a unique solution (meaning, a unique pair of values for x and y) For example:

$$\begin{cases} 4x + 2y = 12 \\ -2x - y = -6 \end{cases}$$

Isolating y in the second equation:

$$y = -2x + 6$$

And using this value in the first equation in the system:

$$\begin{aligned} 4x + 2(-2x + 6) &= 12 \\ 4x - 4x + 12 &= 12 \\ 12 &= 12 \end{aligned}$$

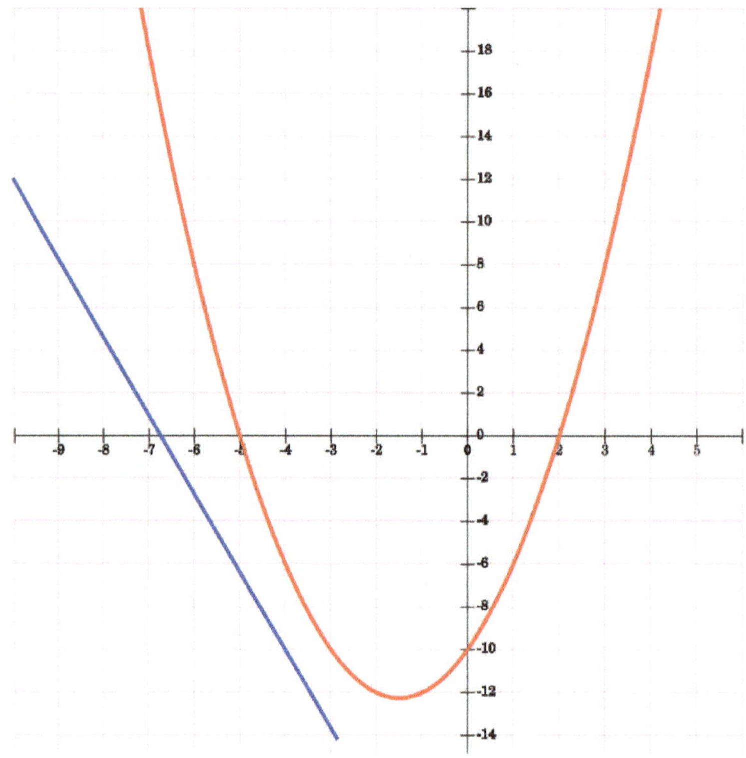

Plot of a quadratic equation (red) and a linear equation (blue) that do not intersect, and consequently for which there is no common solution.

The equality is true, but it does not provide a value for x. Indeed, one can easily verify (by just filling in some values of x that for any x there is a solution as long as $y = -2x + 6$. There is an infinite number of solutions for this system.

Over- and Underdetermined Systems

Systems with more variables than the number of linear equations are called underdetermined. Such a system, if it has any solutions, does not have a unique one but rather an infinitude of them. An example of such a system is

$$\begin{cases} x + 2y = 10 \\ y - z = 2. \end{cases}$$

When trying to solve it, one is led to express some variables as functions of the other ones if any solutions exist, but cannot express *all* solutions numerically because there are an infinite number of them if there are any.

A system with a greater number of equations than variables is called overdetermined. If an overdetermined system has any solutions, necessarily some equations are linear combinations of the others.

REFERENCES

1. Leonhard Euler, Elements of Algebra, 1770. English translation Tarquin Press, 2007, ISBN 978-1-899618-79-8, also online digitized editions 2006, 1822.
2. Charles Smith, A Treatise on Algebra, in Cornell University Library Historical Math Monographs.
3. Redden, John. Elementary Algebra. Flat World Knowledge, 2011
4. H.E. Slaught and N.J. Lennes, Elementary algebra, Publ. Allyn and Bacon, 1915, page 1 (republished by Forgotten Books)
5. Lewis Hirsch, Arthur Goodman, Understanding Elementary Algebra With Geometry: A Course for College Students, Publisher: Cengage Learning, 2005, ISBN 0534999727, 9780534999728, 654 pages, page 2
6. Richard N. Aufmann, Joanne Lockwood, Introductory Algebra: An Applied Approach, Publisher Cengage Learning, 2010, ISBN 1439046042, 9781439046043, page 78
7. William L. Hosch (editor), The Britannica Guide to Algebra and Trigonometry, Britannica Educational Publishing, The Rosen Publishing Group, 2010, ISBN 1615302190, 9781615302192, page 71
8. James E. Gentle, Numerical Linear Algebra for Applications in Statistics, Publisher: Springer, 1998, ISBN 0387985425, 9780387985428, 221 pages, [James E. Gentle page 183]
9. Horatio Nelson Robinson, New elementary algebra: containing the rudiments of science for schools and academies, Ivison, Phinney, Blakeman, & Co., 1866, page 7
10. Ron Larson, Robert Hostetler, Bruce H. Edwards, Algebra And Trigonometry: A Graphing Approach, Publisher: Cengage Learning, 2007, ISBN 061885195X, 9780618851959, 1114 pages, page 6
11. Sin Kwai Meng, Chip Wai Lung, Ng Song Beng, "Algebraic notation", in Mathematics Matters Secondary 1 Express Textbook, Publisher Panpac Education Pte Ltd, ISBN 9812738827, 9789812738820, page 68
12. David Alan Herzog, Teach Yourself Visually Algebra, Publisher John Wiley & Sons, 2008, ISBN 0470185597, 9780470185599, 304 pages, page 72
13. John C. Peterson, Technical Mathematics With Calculus, Publisher Cengage Learning, 2003, ISBN 0766861899, 9780766861893, 1613 pages, page 31
14. Jerome E. Kaufmann, Karen L. Schwitters, Algebra for College Students, Publisher Cengage Learning, 2010, ISBN 0538733543, 9780538733540, 803 pages, page 222
15. Ramesh Bangia, Dictionary of Information Technology, Publisher Laxmi Publications, Ltd., 2010, ISBN 9380298153, 9789380298153, page 212
16. George Grätzer, First Steps in LaTeX, Publisher Springer, 1999, ISBN 0817641327, 9780817641320, page 17

17. S. Tucker Taft, Robert A. Duff, Randall L. Brukardt, Erhard Ploedereder, Pascal Leroy, Ada 2005 Reference Manual, Volume 4348 of Lecture Notes in Computer Science, Publisher Springer, 2007, ISBN 3540693351, 9783540693352, page 13
18. C. Xavier, Fortran 77 And Numerical Methods, Publisher New Age International, 1994, ISBN 812240670X, 9788122406702, page 20
19. Randal Schwartz, Brian Foy, Tom Phoenix, Learning Perl, Publisher O'Reilly Media, Inc., 2011, ISBN 1449313140, 9781449313142, page 24
20. Matthew A. Telles, Python Power!: The Comprehensive Guide, Publisher Course Technology PTR, 2008, ISBN 1598631586, 9781598631586, page 46
21. Kevin C. Baird, Ruby by Example: Concepts and Code, Publisher No Starch Press, 2007, ISBN 1593271484, 9781593271480, page 72
22. William P. Berlinghoff, Fernando Q. Gouvêa, Math through the Ages: A Gentle History for Teachers and Others, Publisher MAA, 2004, ISBN 0883857367, 9780883857366, page 75
23. Thomas Sonnabend, Mathematics for Teachers: An Interactive Approach for Grades K-8, Publisher: Cengage Learning, 2009, ISBN 0495561665, 9780495561668, 759 pages, page xvii
24. Lewis Hirsch, Arthur Goodman, Understanding Elementary Algebra With Geometry: A Course for College Students, Publisher: Cengage Learning, 2005, ISBN 0534999727, 9780534999728, 654 pages, page 48
25. Lawrence S. Leff, College Algebra: Barron's Ez-101 Study Keys, Publisher: Barron's Educational Series, 2005, ISBN 0764129147, 9780764129148, 230 pages, page 2
26. Ron Larson, Kimberly Nolting, Elementary Algebra, Publisher: Cengage Learning, 2009, ISBN 0547102275, 9780547102276, 622 pages, page 210
27. Charles P. McKeague, Elementary Algebra, Publisher: Cengage Learning, 2011, ISBN 0840064217, 9780840064219, 571 pages, page 49
28. Andrew Marx, Shortcut Algebra I: A Quick and Easy Way to Increase Your Algebra I Knowledge and Test Scores, Publisher Kaplan Publishing, 2007, ISBN 1419552880, 9781419552885, 288 pages, page 51
29. Mark Clark, Cynthia Anfinson, Beginning Algebra: Connecting Concepts Through Applications, Publisher Cengage Learning, 2011, ISBN 0534419380, 9780534419387, 793 pages, page 134
30. Alan S. Tussy, R. David Gustafson, Elementary and Intermediate Algebra, Publisher Cengage Learning, 2012, ISBN 1111567689, 9781111567682, 1163 pages, page 493
31. Douglas Downing, Algebra the Easy Way, Publisher Barron's Educational Series, 2003, ISBN 0764119729, 9780764119729, 392 pages, page 20
32. Ron Larson, Robert Hostetler, Intermediate Algebra, Publisher Cengage Learning, 2008, ISBN 0618753524, 9780618753529, 857 pages, page 96
33. http://math.stackexchange.com/a/1043755/19368

34. Chris Carter, Physics: Facts and Practice for A Level, Publisher Oxford University Press, 2001, ISBN 019914768X, 9780199147687, 144 pages, page 50
35. Slavin, Steve (1989). All the Math You'll Ever Need. John Wiley & Sons. p. 72. ISBN 0-471-50636-2.
36. Sinha, The Pearson Guide to Quantitative Aptitude for CAT 2/ePublisher: Pearson Education India, 2010, ISBN 8131723666, 9788131723661, 599 pages, page 195
37. Cynthia Y. Young, Precalculus, Publisher John Wiley & Sons, 2010, ISBN 0471756849, 9780471756842, 1175 pages, page 699
38. Mary Jane Sterling, Algebra II For Dummies, Publisher: John Wiley & Sons, 2006, ISBN 0471775819, 9780471775812, 384 pages, page 37
39. John T. Irwin, The Mystery to a Solution: Poe, Borges, and the Analytic Detective Story, Publisher JHU Press, 1996, ISBN 0801854660, 9780801854668, 512 pages, page 372
40. Sharma/khattar, The Pearson Guide To Objective Mathematics For Engineering Entrance Examinations, 3/E, Publisher Pearson Education India, 2010, ISBN 8131723631, 9788131723630, 1248 pages, page 621
41. Aven Choo, LMAN OL Additional Maths Revision Guide 3, Publisher Pearson Education South Asia, 2007, ISBN 9810600011, 9789810600013, page 105
42. John C. Peterson, Technical Mathematics With Calculus, Publisher Cengage Learning, 2003, ISBN 0766861899, 9780766861893, 1613 pages, page 525
43. Euler's Elements of Algebra
44. Elements of algebra – Leonhard Euler, John Hewlett, Francis Horner, Jean Bernoulli, Joseph Louis Lagrange – Google Books

Number Theory

Chapter Contents

Number theory is a vast and fascinating field of mathematics, sometimes called "higher arithmetic," consisting of the study of the properties of whole numbers. Primes and prime factorization are especially important in number theory, as are a number of functions such as the divisor function, Riemann zeta function, and totient function.

INTRODUCTION

Number theory or, in older usage, arithmetic is a branch of pure mathematics devoted primarily to the study of the integers. It is sometimes called "The Queen of Mathematics" because of its foundational place in the discipline. Number theorists study prime numbers as well as the properties of objects made out of integers (e.g., rational numbers) or defined as generalizations of the integers (e.g., algebraic integers).

Integers can be considered either in themselves or as solutions to equations (Diophantine geometry). Questions in number theory are often best understood through the study of analytical objects (e.g., the Riemann zeta function) that encode properties of the integers, primes or other number-theoretic objects in some fashion (analytic number theory). One may also study real numbers in relation to rational numbers, e.g., as approximated by the latter (Diophantine approximation).

The older term for number theory is arithmetic. By the early twentieth century, it had been superseded by "number theory". (The word "arithmetic" is used by the general public to mean "elementary calculations"; it has also acquired other meanings in mathematical logic, as in Peano arithmetic, and computer science, as in floating point arithmetic.) The use of the term arithmetic for number theory regained some ground in the second half of the 20th century, arguably in part due to French influence. In particular, arithmetical is preferred as an adjective to number-theoretic.

A Lehmer sieve, which is a primitive digital computer once used for finding primes and solving simple Diophantine equations.

Origins

Dawn of Arithmetic

The first historical find of an arithmetical nature is a fragment of a table: the broken clay tablet Plimpton 322 (Larsa, Mesopotamia, ca. 1800 BCE) contains a list of "Pythagorean triples", i.e., integers (a, b, c) such that $a^2 + b^2 = c^2$. The triples are too many and too large to have been obtained by brute force. The heading over the first column reads: "The *takiltum* of the diagonal which has been subtracted such that the width..."

The Plimpton 322 tablet

The table's layout suggests that it was constructed by means of what amounts, in modern language, to the identity

$$\left(\frac{1}{2}\left(x-\frac{1}{x}\right)\right)^2+1=\left(\frac{1}{2}\left(x+\frac{1}{x}\right)\right)^2,$$

which is implicit in routine Old Babylonian exercises. If some other method was used, the triples were first constructed and then reordered by c/a, presumably for actual use as a "table", i.e., with a view to applications.

It is not known what these applications may have been, or whether there could have been any; Babylonian astronomy, for example, truly flowered only later. It has been suggested instead that the table was a source of numerical examples for school problems.

While Babylonian number theory—or what survives of Babylonian mathematics that can be called thus—consists of this single, striking fragment, Babylonian algebra (in the secondary-school sense of "algebra") was exceptionally well developed. Late Neoplatonic sources state that Pythagoras learned mathematics from the Babylonians. Much earlier sources state that Thales and Pythagoras traveled and studied in Egypt.

Euclid IX 21—34 is very probably Pythagorean; it is very simple material ("odd times even is even", "if an odd number measures [= divides] an even number, then it also measures [= divides] half of it"), but it is all that is needed to prove that $\sqrt{2}$ is irrational. Pythagorean mystics gave great importance to the odd and the even. The discovery that $\sqrt{2}$ is irrational is credited to the early Pythagoreans (pre-Theodorus). By revealing (in modern terms) that numbers could be irrational, this discovery seems to have provoked the first foundational crisis in mathematical history; its proof or its divulgation are sometimes credited to Hippasus, who was expelled or split from the Pythagorean sect. This forced a distinction between *numbers* (integers and the rationals—the subjects of arithmetic), on the one hand, and *lengths* and *proportions* (which we would identify with real numbers, whether rational or not), on the other hand.

The Pythagorean tradition spoke also of so-called polygonal or figurate numbers. While square numbers, cubic numbers, etc., are seen now as more natural than triangular numbers, pentagonal numbers, etc., the study of the sums of triangular and pentagonal numbers would prove fruitful in the early modern period (17th to early 19th century).

We know of no clearly arithmetical material in ancient Egyptian or Vedic sources, though there is some algebra in both. The Chinese remainder theorem appears as an exercise in Sun Zi's *Suan Ching*, also known as *The Mathematical Classic of Sun Zi* (3rd, 4th or 5th century CE.) (There is one important step glossed over in Sun Zi's solution:it is the problem that was later solved by Āryabhata's Kuttaka – see below.)

There is also some numerical mysticism in Chinese mathematics, but, unlike that of the Pythagoreans, it seems to have led nowhere. Like the Pythagoreans' perfect numbers, magic squares have passed from superstition into recreation.

Classical Greece and the Early Hellenistic Period

Aside from a few fragments, the mathematics of Classical Greece is known to us either through the reports of contemporary non-mathematicians or through mathematical works from the early Hellenistic period. In the case of number theory, this means, by and large, *Plato* and *Euclid*, respectively.

Plato had a keen interest in mathematics, and distinguished clearly between arithmetic and calculation. (By *arithmetic* he meant, in part, theorising on number, rather than what *arithmetic* or *number theory* have come to mean.) It is through one of Plato's dialogues—namely, *Theaetetus*—that we know that Theodorus had proven that $\sqrt{3}, \sqrt{5}, \ldots, \sqrt{17}$ are irrational. Theaetetus was, like Plato, a disciple of Theodorus's; he worked on distinguishing different kinds of incommensurables, and was thus arguably a pioneer in the study of number systems. (Book X of Euclid's Elements is described by Pappus as being largely based on Theaetetus's work.)

Euclid devoted part of his *Elements* to prime numbers and divisibility, topics that belong unambiguously to number theory and are basic to it (Books VII to IX of Euclid's Elements). In particular, he gave an algorithm for computing the greatest common divisor of two numbers (the Euclidean algorithm; *Elements*, Prop. VII.2) and the first known proof of the infinitude of primes (*Elements*, Prop. IX.20).

In 1773, Lessing published an epigram he had found in a manuscript during his work as a librarian; it claimed to be a letter sent by Archimedes to Eratosthenes. The epigram proposed what has become known as Archimedes' cattle problem; its solution (absent from the manuscript) requires solving an indeterminate quadratic equation (which reduces to what would later be misnamed Pell's equation). As far as we know, such equations were first successfully treated by the Indian school. It is not known whether Archimedes himself had a method of solution.

Diophantus

Title page of the 1621 edition of Diophantus' Arithmetica, translated into Latin by Claude Gaspard Bachet de Méziriac.

Very little is known about Diophantus of Alexandria; he probably lived in the third century CE, that is, about five hundred years after Euclid. Six out of the thirteen books of Diophantus's *Arithmetica* survive in the original Greek; four more books survive in an Arabic translation. The *Arithmetica* is a collection of worked-out problems where the task is invariably to find rational solutions to a system of polynomial equations, usually of the form $f(x,y) = z^2$ or $f(x,y,z) = w^2$. Thus, nowadays, we speak of *Diophantine equations* when we speak of polynomial equations to which rational or integer solutions must be found.

One may say that Diophantus was studying rational points — i.e., points whose coordinates are rational — on curves and algebraic varieties; however, unlike the Greeks of the Classical period, who did what we would now call basic algebra in geometrical terms, Diophantus did what we would now call basic algebraic geometry in purely algebraic terms. In modern language, what Diophantus did was to find rational parametrizations of varieties; that is, given an equation of the form (say) $f(x_1, x_2, x_3) = 0$, his aim was to find (in essence) three rational functions g_1, g_2, g_3 such that, for all values of r and s, setting $x_i = g_i(r,s)$ $i = 1,2,3$ gives a solution to $f(x_1, x_2, x_3) = 0$.

Diophantus also studied the equations of some non-rational curves, for which no rational parametrisation is possible. He managed to find some rational points on these curves (elliptic curves, as it happens, in what seems to be their first known occurrence) by means of what amounts to a tangent construction: translated into coordinate geometry (which did not exist in Diophantus's time), his method would be visualised as drawing a tangent to a curve at a known rational point, and then finding the other point of intersection of the tangent with the curve; that other point is a new rational point. (Diophantus also resorted to what could be called a special case of a secant construction.)

While Diophantus was concerned largely with rational solutions, he assumed some results on integer numbers, in particular that every integer is the sum of four squares (though he never stated as much explicitly).

Āryabhata, Brahmagupta, Bhāskara

While Greek astronomy probably influenced Indian learning, to the point of introducing trigonometry, it seems to be the case that Indian mathematics is otherwise an indigenous tradition; in particular, there is no evidence that Euclid's Elements reached India before the 18th century.

Āryabhata (476–550 CE) showed that pairs of simultaneous congruences {\displaystyle n\equiv a $n \equiv a_1 \bmod m_1$, $n \equiv a_2 \bmod m_2$ could be solved by a method he called *kuttaka*, or *pulveriser*; this is a procedure close to (a generalisation of) the Euclidean algorithm, which was probably discovered independently in India. Āryabhata seems to have had in mind applications to astronomical calculations.

Brahmagupta (628 CE) started the systematic study of indefinite quadratic equations—in particular, the misnamed Pell equation, in which Archimedes may have first been interested, and which did not start to be solved in the West until the time of Fermat and Euler. Later Sanskrit authors would follow, using Brahmagupta's technical terminology. A general procedure (the chakravala, or "cyclic method") for solving Pell's equation was finally found by Jayadeva (cited in the eleventh century; his work is otherwise lost); the earliest surviving exposition appears in Bhāskara II's Bīja-ganita (twelfth century).

Indian mathematics remained largely unknown in Europe until the late eighteenth century; Brahmagupta and Bhāskara's work was translated into English in 1817 by Henry Colebrooke.

Arithmetic in the Islamic Golden Age

In the early ninth century, the caliph Al-Ma'mun ordered translations of many Greek mathematical works and at least one Sanskrit work (the *Sindhind*, which may or may not be Brahmagupta's Brāhmasphuṭasiddhānta). Diophantus's main work, the *Arithmetica*, was translated into Arabic by Qusta ibn Luqa (820–912). Part of the treatise *al-Fakhri* (by al-Karajī, 953 – ca. 1029) builds on it to some extent. According to Rashed Roshdi, Al-Karajī's contemporary Ibn al-Haytham knew what would later be called Wilson's theorem.

Western Europe in the Middle Ages

Al-Haytham seen by the West: frontispice of *Selenographia*, showing Alhasen [*sic*] representing knowledge through reason, and Galileo representing knowledge through the senses.

Early Modern Number Theory

Fermat

Pierre de Fermat

Pierre de Fermat (1601–1665) never published his writings; in particular, his work on number theory is contained almost entirely in letters to mathematicians and in private marginal notes. He wrote down nearly no proofs in number theory; he had no models in the area.He did make repeated use of mathematical induction, introducing the method of infinite descent.

One of Fermat's first interests was perfect numbers (which appear in Euclid, *Elements* IX) and amicable numbers; this led him to work on integer divisors, which were from the beginning among the subjects of the correspondence (1636 onwards) that put him in touch with the mathematical community of the day. He had already studied Bachet's edition of Diophantus carefully; by 1643, his interests had shifted largely to Diophantine problems and sums of squares (also treated by Diophantus).

Fermat's achievements in arithmetic include:

Fermat's little theorem (1640), stating that, if a is not divisible by a prime p, then $a^{p-1} \equiv 1 \bmod p$.

- Fermat's little theorem (1640), stating that, if a is not divisible by a prime p, then $a^{p-1} \equiv 1 \bmod p$.
- If a and b are coprime, then $a^2 + b^2$ is not divisible by any prime congruent to −1 modulo 4; and every prime congruent to 1 modulo 4 can be written in the form $a^2 + b^2$. These two statements also date from 1640; in 1659, Fermat stated to Huygens that he had proven the latter statement by the method of

infinite descent. Fermat and Frenicle also did some work (some of it erroneous) on other quadratic forms.

- Fermat posed the problem of solving $x^2 - Ny^2 = 1$ as a challenge to English mathematicians (1657). The problem was solved in a few months by Wallis and Brouncker. Fermat considered their solution valid, but pointed out they had provided an algorithm without a proof (as had Jayadeva and Bhaskara, though Fermat would never know this.) He states that a proof can be found by descent.
- Fermat developed methods for (doing what in our terms amounts to) finding points on curves of genus 0 and 1. As in Diophantus, there are many special procedures and what amounts to a tangent construction, but no use of a secant construction.
- Fermat states and proves (by descent) in the appendix to Observations on Diophantus (Obs. XLV) that $x^4 + y^4 = z^4$ has no non-trivial solutions in the integers. Fermat also mentioned to his correspondents that $x^3 + y^3 = z^3$ has no non-trivial solutions, and that this could be proven by descent. The first known proof is due to Euler (1753; indeed by descent).

Fermat's claim ("Fermat's last theorem") to have shown there are no solutions to $x^n + y^n = z^n$ for all $n \geq 3$ (the only known proof of which is beyond his methods) appears only in his annotations on the margin of his copy of Diophantus; he never claimed this to others and thus would have had no need to retract it if he found any mistake in his supposed proof.

Euler

Leonhard Euler

The interest of Leonhard Euler (1707–1783) in number theory was first spurred in 1729, when a friend of his, the amateur[note 8] Goldbach, pointed him towards some of Fermat's work on the subject. This has been called the "rebirth" of modern number theory, after Fermat's relative lack of success in getting his contemporaries' attention for the subject.Euler's work on number theory includes the following:

- Proofs for Fermat's statements. This includes Fermat's little theorem (generalised by Euler to non-prime moduli); the fact that $p = x^2 + y^2$ if and only if $p \equiv 1 \bmod 4$; initial work towards a proof that every integer is the sum of four squares (the first complete proof is by Joseph-Louis Lagrange (1770), soon improved by Euler himself); the lack of non-zero integer solutions to $x^4 + y^4 = z^2$ (implying the case n=4 of Fermat's last theorem, the case n=3 of which Euler also proved by a related method).
- *Pell's equation*, first misnamed by Euler. He wrote on the link between continued fractions and Pell's equation.
- *First steps towards analytic number theory.* In his work of sums of four squares, partitions, pentagonal numbers, and the distribution of prime numbers, Euler pioneered the use of what can be seen as analysis (in particular, infinite series) in number theory. Since he lived before the development of complex analysis, most of his work is restricted to the formal manipulation of power series. He did, however, do some very notable (though not fully rigorous) early work on what would later be called the Riemann zeta function.
- *Quadratic forms.* Following Fermat's lead, Euler did further research on the question of which primes can be expressed in the form $x^2 + Ny^2$, some of it prefiguring quadratic reciprocity.
- *Diophantine equations.* Euler worked on some Diophantine equations of genus 0 and 1. In particular, he studied Diophantus's work; he tried to systematise it, but the time was not yet ripe for such an endeavour – algebraic geometry was still in its infancy. He did notice there was a connection between Diophantine problems and elliptic integrals, whose study he had himself initiated.

Lagrange, Legendre and Gauss

Joseph-Louis Lagrange (1736–1813) was the first to give full proofs of some of Fermat's and Euler's work and observations – for instance, the four-square theorem and the basic theory of the misnamed "Pell's equation" (for which an algorithmic solution was found by Fermat and his contemporaries, and also by Jayadeva and Bhaskara II before them.) He also studied quadratic forms in full generality (as opposed to $mX^2 + nY^2$) — defining their equivalence relation, showing how to put them in reduced form, etc.

Adrien-Marie Legendre (1752–1833) was the first to state the law of quadratic reciprocity. He also conjectured what amounts to the prime number theorem and Dirichlet's theorem on arithmetic progressions. He gave a full treatment of the equation $ax^2 + by^2 + cz^2 = 0$ and worked on quadratic forms along the lines later developed fully by Gauss. In his old age, he was the first to prove "Fermat's last theorem" for n=5 (completing work by Peter Gustav Lejeune Dirichlet, and crediting both him and Sophie Germain).

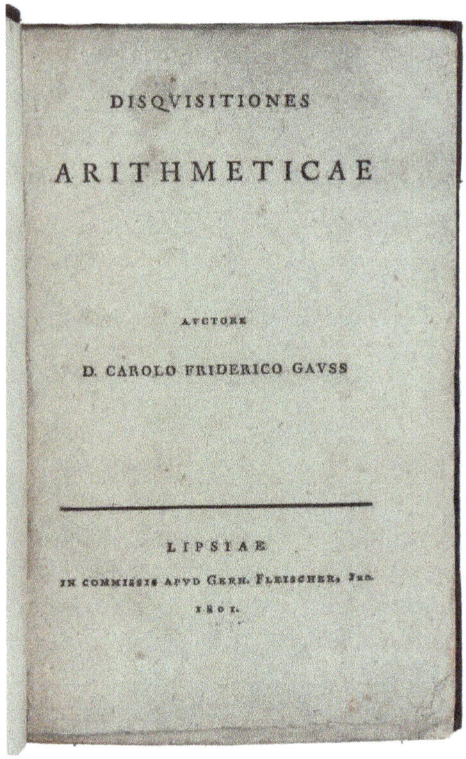

DISQVISITIONES

ARITHMETICAE

AVCTORE

D. CAROLO FRIDERICO GAVSS

LIPSIAE

IN COMMISSIS APVD GERH. FLEISCHER, IUN.

1801.

Carl Friedrich Gauss's Disquisitiones Arithmeticae, first edition

Carl Friedrich Gauss

The theory of the division of the circle...which is treated in sec. 7 does not belong by itself to arithmetic, but its principles can only be drawn from higher arithmetic.

In this way, Gauss arguably made a first foray towards both Évariste Galois's work and algebraic number theory.

Maturity and Division Into Subfields

Starting early in the nineteenth century, the following developments gradually took place:

- The rise to self-consciousness of number theory (or *higher arithmetic*) as a field of study.
- The development of much of modern mathematics necessary for basic modern number theory: complex analysis, group theory, Galois theory—accompanied by greater rigor in analysis and abstraction in algebra.
- The rough subdivision of number theory into its modern subfields—in particular, analytic and algebraic number theory.

Algebraic number theory may be said to start with the study of reciprocity and cyclotomy, but truly came into its own with the development of abstract algebra and early ideal theory and valuation theory; see below. A conventional starting point for analytic number theory is Dirichlet's theorem on arithmetic progressions (1837), whose proof introduced L-functions and involved some asymptotic analysis and a limiting process on a real variable. The first use of analytic ideas in number theory actually goes back to Euler (1730s), who used formal power series and non-rigorous (or implicit) limiting arguments. The use of *complex* analysis in number theory comes later: the work of Bernhard Riemann (1859) on the zeta function is the canonical starting point; Jacobi's four-square theorem (1839), which predates it, belongs to an initially different strand that has by now taken a leading role in analytic number theory (modular forms).

The history of each subfield is briefly addressed in its own section below; see the main article of each subfield for fuller treatments. Many of the most interesting questions in each area remain open and are being actively worked on.

Ernst Kummer

Peter Gustav Lejeune Dirichlet

MAIN SUBDIVISIONS

Elementary Tools

The term *elementary* generally denotes a method that does not use complex analysis. For example, the prime number theorem was first proven using complex analysis in 1896, but an elementary proof was found only in 1949 by Erdős and Selberg. The term is somewhat ambiguous: for example, proofs based on complex Tauberian theorems (e.g. Wiener–Ikehara) are often seen as quite enlightening but not elementary, in spite of using Fourier analysis, rather than complex analysis as such. Here as elsewhere, an *elementary* proof may be longer and more difficult for most readers than a non-elementary one.

Number theory has the reputation of being a field many of whose results can be stated to the layperson. At the same time, the proofs of these results are not particularly accessible, in part because the range of tools they use is, if anything, unusually broad within mathematics.

Analytic Number Theory

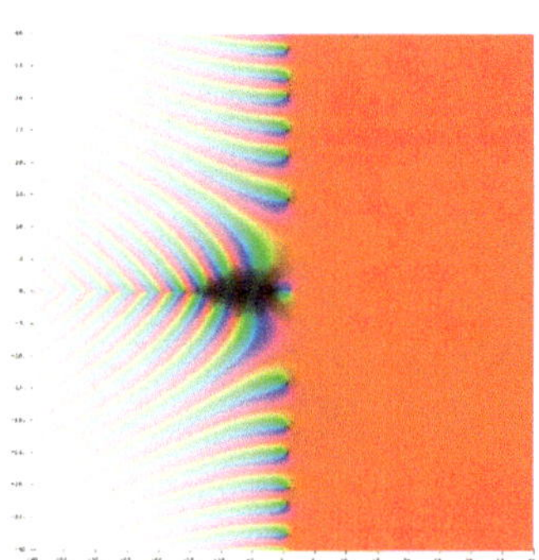

Riemann zeta function $\zeta(s)$ in the complex plane. The color of a point s gives the value of $\zeta(s)$: dark colors denote values close to zero and hue gives the value's argument.

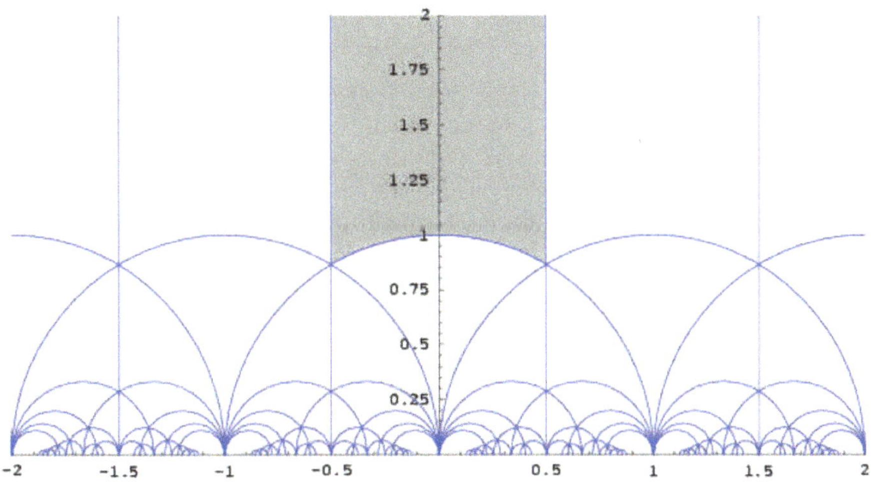

The action of the modular group on the upper half plane. The region in grey is the standard fundamental domain.

Analytic number theory ***may be defined***

- in terms of its tools, as the study of the integers by means of tools from real and complex analysis; or
- in terms of its concerns, as the study within number theory of estimates on size and density, as opposed to identities.

Some subjects generally considered to be part of analytic number theory, e.g., sieve theory,[note 9] are better covered by the second rather than the first definition: some of sieve theory, for instance, uses little analysis,[note 10] yet it does belong to analytic number theory.

The following are examples of problems in analytic number theory: the prime number theorem, the Goldbach conjecture (or the twin prime conjecture, or the Hardy–Littlewood conjectures), the Waring problem and the Riemann hypothesis. Some of the most important tools of analytic number theory are the circle method, sieve methods and L-functions (or, rather, the study of their properties). The theory of modular forms (and, more generally, automorphic forms) also occupies an increasingly central place in the toolbox of analytic number theory.

One may ask analytic questions about algebraic numbers, and use analytic means to answer such questions; it is thus that algebraic and analytic number theory intersect. For example, one may define prime ideals (generalizations of prime numbers in the field of algebraic numbers) and ask how many prime ideals there are up to a certain size. This question can be answered by means of an examination of Dedekind zeta functions, which are generalizations of the Riemann zeta function, a key analytic object at the roots of the subject. This is an example of a general procedure in analytic number theory: deriving information about the distribution of a sequence (here, prime ideals or prime numbers) from the analytic behavior of an appropriately constructed complex-valued function.

Algebraic Number Theory

An *algebraic number* is any complex number that is a solution to some polynomial equation $f(x) = 0$ with rational coefficients; for example, every solution $x^5 + (11/2)x^3 - 7x^2 + 9 = 0$ (say) is an algebraic number. Fields of algebraic numbers are also called *algebraic number fields*, or shortly *number fields*. Algebraic number theory studies algebraic number fields. Thus, analytic and algebraic number theory can and do overlap: the former is defined by its methods, the latter by its objects of study.

It could be argued that the simplest kind of number fields (viz., quadratic fields) were already studied by Gauss, as the discussion of quadratic forms in *Disquisitiones arithmeticae* can be restated in terms of ideals and norms in quadratic fields. (A *quadratic field* consists of all numbers of the form $a + b\sqrt{d}$, where a and b are rational numbers and d is a fixed rational number whose square root is not rational.) For that matter, the 11th-century chakravala method amounts—in modern terms—to an algorithm for finding the units of a real quadratic number field. However, neither Bhāskara nor Gauss knew of number fields as such.

The grounds of the subject as we know it were set in the late nineteenth century, when *ideal numbers*, the *theory of ideals* and *valuation theory* were developed; these are three complementary ways of dealing with the lack of unique factorisation in algebraic number fields. (For example, in the field generated by the rationals and $\sqrt{-5}$, the number 6 can be factorised both as $6 = 2 \cdot 3$ and $6 = (1 + \sqrt{-5})(1 - \sqrt{-5})$; all of 2, 3, $1 + \sqrt{-5}$ and $1 - \sqrt{-5}$ are irreducible, and thus, in a naïve sense, analogous to primes among the integers.) The initial impetus for the development of ideal numbers (by Kummer) seems to have come from the study of higher reciprocity laws,i.e., generalisations of quadratic reciprocity.

Number fields are often studied as extensions of smaller number fields: a field *L* is said to be an *extension* of a field *K* if *L* contains *K*. (For example, the complex numbers *C* are an extension of the reals *R*, and the reals *R* are an extension of the rationals *Q*.) Classifying the possible extensions of a given number field is a difficult and partially open problem. Abelian extensions—that is, extensions *L* of *K* such that the Galois group[note 11] Gal(*L*/*K*) of *L* over *K* is an abelian group—are relatively well understood. Their classification was the object of the programme of class field theory, which was initiated in the late 19th century (partly by Kronecker and Eisenstein) and carried out largely in 1900—1950.

An example of an active area of research in algebraic number theory is Iwasawa theory. The Langlands program, one of the main current large-scale research plans in mathematics, is sometimes described as an attempt to generalise class field theory to non-abelian extensions of number fields.

Diophantine Geometry

The central problem of *Diophantine geometry* is to determine when a Diophantine equation has solutions, and if it does, how many. The approach taken is to think of the solutions of an equation as a geometric object.

For example, an equation in two variables defines a curve in the plane. More generally, an equation, or system of equations, in two or more variables defines a curve, a surface or some other such object in *n*-dimensional space. In Diophantine geometry, one asks whether there are any *rational points* (points all of whose coordinates are rationals) or *integral points* (points all of whose coordinates are integers) on the curve or surface. If there are any such points, the next step is to ask how many there are and how they are distributed. A basic question in this direction is: are there finitely or infinitely many rational points on a given curve (or surface)? What about integer points?

An example here may be helpful. Consider the Pythagorean equation $x^2 + y^2 = 1$; we would like to study its rational solutions, i.e., its solutions (x, y) such that x and y are both rational. This is the same as asking for all integer solutions to $a^2 + b^2 = c^2$; any solution to the latter equation gives us a solution $x = a/c, y = b/c$ to the former. It is also the same as asking for all points with rational coordinates on the curve described by $x^2 + y^2 = 1$. (This curve happens to be a circle of radius 1 around the origin.)

The rephrasing of questions on equations in terms of points on curves turns out to be felicitous. The finiteness or not of the number of rational or integer points on an algebraic curve—that is, rational or integer solutions to an equation $f(x, y) = 0$, where f is a polynomial in two variables—turns out to depend crucially on the genus of the curve. The genus can be defined as follows: allow the variables in f (x,y) = 0 to be complex numbers; then f (x,y) = 0 defines a 2-dimensional surface in (projective) 4-dimensional space (since two complex variables can be decomposed into four real variables, i.e., four dimensions). Count the number of (doughnut) holes in the surface; call this number the genus of f (x,y) = 0 . Other geometrical notions turn out to be just as crucial.

There is also the closely linked area of Diophantine approximations: given a number x how well can it be approximated by rationals? (We are looking for approximations that are good relative to the amount of space that it takes to write the rational: call a/q (with gcd(a,q) = 1) a good approximation to x if $|x - a/q| < \frac{1}{q^c}$, where c is large.) This question is of special interest if x is an algebraic number. If x cannot be well approximated, then some equations do not have integer or rational solutions. Moreover, several concepts (especially that of height) turn out to be crucial both in Diophantine geometry and in the study of Diophantine approximations. This question is also of special interest in transcendental number theory: if a number can be better approximated than any algebraic number, then it is a transcendental number. It is by this argument that π and e have been shown to be transcendental.

Diophantine geometry should not be confused with the geometry of numbers, which is a collection of graphical methods for answering certain questions in algebraic number theory. *Arithmetic geometry*, on the other hand, is a contemporary term for much the same domain as that covered by the term *Diophantine geometry*. The term *arithmetic geometry* is arguably used most often when one wishes to emphasise the connections to modern algebraic geometry (as in, for instance, Faltings' theorem) rather than to techniques in Diophantine approximations.

Recent Approaches and Subfields

The areas below date as such from no earlier than the mid-twentieth century, even if they are based on older material. For example, as is explained below, the matter of algorithms in number theory is very old, in some sense older than the concept of proof; at the same time, the modern study of computability dates only from the 1930s and 1940s, and computational complexity theory from the 1970s.

Probabilistic Number Theory

Take a number at random between one and a million. How likely is it to be prime? This is just another way of asking how many primes there are between one and a million. Further: how many prime divisors will it have, on average? How many divisors will it have altogether, and with what likelihood? What is the probability that it will have many more or many fewer divisors or prime divisors than the average?

Much of probabilistic number theory can be seen as an important special case of the study of variables that are almost, but not quite, mutually independent. For example, the event that a random integer between one and a million be divisible by two and the event that it be divisible by three are almost independent, but not quite.

It is sometimes said that probabilistic combinatorics uses the fact that whatever happens with probability greater than 0 must happen sometimes; one may say with equal justice that many applications of probabilistic number theory hinge on the fact that whatever is unusual must be rare. If certain algebraic objects (say, rational or integer solutions to certain equations) can be shown to be in the tail of certain sensibly defined distributions, it follows that there must be few of them; this is a very concrete non-probabilistic statement following from a probabilistic one.

At times, a non-rigorous, probabilistic approach leads to a number of heuristic algorithms and open problems, notably Cramér's conjecture.

Arithmetic Combinatorics

Let A be a set of N integers. Consider the set $A + A = \{ m + n \mid m, n \in A \}$ consisting of all sums of two elements of A. Is $A + A$ much larger than A? Barely larger? If $A + A$ is barely larger than A, must A have plenty of arithmetic structure, for example, does A resemble an arithmetic progression?

If we begin from a fairly "thick" infinite set A, does it contain many elements in arithmetic progression: a, $a+b, a+2b, a+3b, \ldots, a+10b$, say? Should it be possible to write large integers as sums of elements of A?

These questions are characteristic of *arithmetic combinatorics*. This is a presently coalescing field; it subsumes *additive number theory* (which concerns itself with certain very specific sets A of arithmetic significance, such as the primes or the squares) and, arguably, some of the *geometry of numbers*, together with some rapidly developing new material. Its focus on issues of growth and distribution accounts in part for its developing links with ergodic theory, finite group theory, model theory, and other fields. The term *additive combinatorics* is also used; however, the sets A being studied need not be sets of integers, but rather subsets of non-commutative groups, for which the multiplication symbol, not the addition symbol, is traditionally used; they can also be subsets of rings, in which case the growth of $A+A$ and $A \cdot A$ may be compared.

Computations in Number Theory

While the word *algorithm* goes back only to certain readers of al-Khwārizmī, careful descriptions of methods of solution are older than proofs: such methods (that is, algorithms) are as old as any recognisable mathematics—ancient Egyptian, Babylonian, Vedic, Chinese—whereas proofs appeared only with the Greeks of the classical period. An interesting early case is that of what we now call the Euclidean algorithm. In its basic form (namely, as an algorithm for computing the greatest common divisor) it appears as Proposition 2 of Book VII in *Elements*, together with a proof of correctness. However, in the form that is often used in number theory (namely, as an algorithm for finding integer solutions to an equation $ax + by = c$, or, what is the same, for finding the quantities whose existence is assured by the Chinese remainder theorem) it first appears in the works of Āryabhata (5th 6th century CE) as an algorithm called *kuttaka* ("pulveriser"), without a proof of correctness.

There are two main questions: "can we compute this?" and "can we compute it rapidly?". Anybody can test whether a number is prime or, if it is not, split it into prime factors; doing so rapidly is another matter. We now know fast algorithms for testing primality, but, in spite of much work (both theoretical and practical), no truly fast algorithm for factoring.

The difficulty of a computation can be useful: modern protocols for encrypting messages (e.g., RSA) depend on functions that are known to all, but whose inverses (a) are known only to a chosen few, and (b) would take one too long a time to figure out on one's own. For example, these functions can be such that their inverses can be computed only if certain large integers are factorized. While many difficult computational problems outside number theory are known, most working encryption protocols nowadays are based on the difficulty of a few number-theoretical problems.

On a different note — some things may not be computable at all; in fact, this can be proven in some instances. For instance, in 1970, it was proven, as a solution to Hilbert's 10th problem, that there is no Turing machine which can solve all Diophantine equations. In particular, this means that, given a computably enumerable set of axioms, there are Diophantine equations for which there is no proof, starting from the axioms, of whether the set of equations has or does not have integer solutions. (We would necessarily be speaking of Diophantine equations for which there are no integer solutions, since, given a Diophantine equation with at least one solution, the solution itself provides a proof of the fact that a solution exists. We cannot prove, of course, that a particular Diophantine equation is of this kind, since this would imply that it has no solutions.)

APPLICATIONS

The number-theorist Leonard Dickson (1874–1954) said "Thank God that number theory is unsullied by any application". Such a view is no longer applicable to number theory. In 1974, Donald Knuth said "...virtually every theorem in elementary number theory arises in a natural, motivated way in connection with the problem of making computers do high-speed numerical calculations". Elementary number theory is taught in discrete mathematics courses for computer scientists; and, on the other hand, number theory also has applications to the continuous in numerical analysis. As well as the well-known applications to cryptography, there are also applications to many other areas of mathematics

Literature

Two of the most popular introductions to the subject are:

- G. H. Hardy; E. M. Wright (2008) [1938]. An introduction to the theory of numbers (rev. by D. R. Heath-Brown and J. H. Silverman, 6th ed.). Oxford University Press. ISBN 978-0-19-921986-5. Retrieved 2016-03-02.
- Vinogradov, I. M. (2003) [1954]. Elements of Number Theory (reprint of the 1954 ed.). Mineola, NY: Dover Publications.

Hardy and Wright's book is a comprehensive classic, though its clarity sometimes suffers due to the authors' insistence on elementary methods. Vinogradov's main attraction consists in its set of problems, which quickly lead to Vinogradov's own research interests; the text itself is very basic and close to minimal. Other popular first introductions are:

- Ivan M. Niven; Herbert S. Zuckerman; Hugh L. Montgomery (2008) [1960]. An introduction to the theory of numbers(reprint of the 5th edition 1991 ed.). John Wiley & Sons. ISBN 978-81-265-1811-1. Retrieved 2016-02-28.
- Kenneth H. Rosen (2010). Elementary Number Theory (6th ed.). Pearson Education. ISBN 978-0-321-71775-7. Retrieved 2016-02-28.

Popular choices for a second textbook include:

- Borevich, A. I.; Shafarevich, Igor R. (1966). Number theory. Pure and Applied Mathematics. **20**. Boston, MA: Academic Press. ISBN 978-0-12-117850-5. MR 0195803.
- Serre, Jean-Pierre (1996) [1973]. A course in arithmetic. Graduate texts in mathematics. 7. Springer. ISBN 978-0-387-90040-7.

REFERENCES

1. Long 1972, p. 1.
2. Neugebauer & Sachs 1945, p. 40. The term takiltum is problematic. Robson prefers the rendering "The holding-square of the diagonal from which 1 is torn out, so that the short side comes up...".Robson 2001, p. 192
3. Robson 2001, p. 189. Other sources give the modern formula . Van der Waerden gives both the modern formula and what amounts to the form preferred by Robson.(van der Waerden 1961, p. 79)
4. van der Waerden 1961, p. 184.
5. Neugebauer (Neugebauer 1969, pp. 36–40) discusses the table in detail and mentions in passing Euclid's method in modern notation (Neugebauer 1969, p. 39).
6. Friberg 1981, p. 302.
7. van der Waerden 1961, p. 43.
8. Iamblichus, Life of Pythagoras,(trans. e.g. Guthrie 1987) cited in van der Waerden 1961, p. 108. See also Porphyry, Life of Pythagoras, paragraph 6, in Guthrie

1987 Van der Waerden (van der Waerden 1961, pp. 87–90) sustains the view that Thales knew Babylonian mathematics.

9. Herodotus (II. 81) and Isocrates (Busiris 28), cited in: Huffman 2011. On Thales, see Eudemus ap. Proclus, 65.7, (e.g. Morrow 1992, p. 52) cited in: O'Grady 2004, p. 1. Proclus was using a work by Eudemus of Rhodes (now lost), the Catalogue of Geometers. See also introduction, Morrow 1992, p. xxx on Proclus' reliability.
10. Becker 1936, p. 533, cited in: van der Waerden 1961, p. 108.
11. Becker 1936.
12. van der Waerden 1961, p. 109.
13. Plato, Theaetetus, p. 147 B, (e.g. Jowett 1871), cited in von Fritz 2004, p. 212: "Theodorus was writing out for us something about roots, such as the roots of three or five, showing that they are incommensurable by the unit;..." See also Spiral of Theodorus.
14. von Fritz 2004.
15. Heath 1921, p. 76.
16. Sun Zi, Suan Ching, Chapter 3, Problem 26. This can be found in Lam & Ang 2004, pp. 219–220, which contains a full translation of the Suan Ching (based on Qian 1963). See also the discussion in Lam & Ang 2004, pp. 138–140.
17. The date of the text has been narrowed down to 220–420 AD (Yan Dunjie) or 280–473 AD (Wang Ling) through internal evidence (= taxation systems assumed in the text). See Lam & Ang 2004, pp. 27–28.
18. Boyer & Merzbach 1991, p. 82.
19. Vardi 1998, p. 305–319.
20. Weil 1984, pp. 17–24.
21. Plofker 2008, p. 119.
22. Any early contact between Babylonian and Indian mathematics remains conjectural (Plofker 2008, p. 42).
23. Mumford 2010, p. 387.
24. Āryabhata, Āryabhatīya, Chapter 2, verses 32–33, cited in: Plofker 2008, pp. 134–140. See also Clark 1930, pp. 42–50. A slightly more explicit description of the kuttaka was later given in Brahmagupta, Brāhmasphutasiddhānta, XVIII, 3–5 (in Colebrooke 1817, p. 325, cited in Clark 1930, p. 42).
25. Mumford 2010, p. 388.
26. Plofker 2008, p. 194.
27. Plofker 2008, p. 283.
28. Colebrooke 1817.
29. Colebrooke 1817, p. lxv, cited in Hopkins 1990, p. 302. See also the preface in Sachau 1888 cited in Smith 1958, pp. 168

30. Pingree 1968, pp. 97–125, and Pingree 1970, pp. 103–123, cited in Plofker 2008, p. 256.
31. Rashed 1980, p. 305–321.
32. Weil 1984, pp. 45–46.
33. Weil 1984, p. 118. This was more so in number theory than in other areas (remark in Mahoney 1994, p. 284). Bachet's own proofs were "ludicrously clumsy" (Weil 1984, p. 33).
34. Mahoney 1994, pp. 48, 53–54. The initial subjects of Fermat's correspondence included divisors ("aliquot parts") and many subjects outside number theory; see the list in the letter from Fermat to Roberval, 22.IX.1636, Tannery & Henry 1891, Vol. II, pp. 72, 74, cited in Mahoney 1994, p. 54.
35. Weil 1984, pp. 1–2.
36. Weil 1984, p. 53.
37. Tannery & Henry 1891, Vol. II, p. 209, Letter XLVI from Fermat to Frenicle, 1640, cited in Weil 1984, p. 56
38. Tannery & Henry 1891, Vol. II, p. 204, cited in Weil 1984, p. 63. All of the following citations from Fermat's Varia Opera are taken from Weil 1984, Chap. II. The standard Tannery & Henry work includes a revision of Fermat's posthumous Varia Opera Mathematica originally prepared by his son (Fermat 1679).
39. Tannery & Henry 1891, Vol. II, p. 213.
40. Tannery & Henry 1891, Vol. II, p. 423.
41. Weil 1984, pp. 80, 91–92.
42. Weil 1984, p. 92.
43. Weil 1984, Ch. II, sect. XV and XVI.
44. Tannery & Henry 1891, Vol. I, pp. 340–341.
45. Weil 1984, p. 115.
46. Weil 1984, pp. 115–116.
47. Weil 1984, p. 104.
48. Weil 1984, pp. 2, 172.
49. Varadarajan 2006, p. 9.
50. Weil 1984, p. 2 and Varadarajan 2006, p. 37
51. Varadarajan 2006, p. 39 and Weil 1984, pp. 176–189
52. Weil 1984, pp. 178–179.
53. Weil 1984, p. 174. Euler was generous in giving credit to others (Varadarajan 2006, p. 14), not always correctly.
54. Weil 1984, p. 183.
55. Varadarajan 2006, pp. 45–55; see also chapter III.
56. Varadarajan 2006, pp. 44–47.

57. Weil 1984, pp. 177–179.
58. Edwards 1983, pp. 285–291.
59. Varadarajan 2006, pp. 55–56.
60. Weil 1984, pp. 179–181.
61. Weil 1984, p. 181.
62. Weil 1984, pp. 327–328.
63. Weil 1984, pp. 332–334.
64. Weil 1984, pp. 337–338.
65. Goldstein & Schappacher 2007, p. 14.
66. From the preface of Disquisitiones Arithmeticae; the translation is taken from Goldstein & Schappacher 2007, p. 16
67. See the discussion in section 5 of Goldstein & Schappacher 2007. Early signs of self-consciousness are present already in letters by Fermat: thus his remarks on what number theory is, and how "Diophantus's work [...] does not really belong to [it]" (quoted in Weil 1984, p. 25).
68. Apostol 1976, p. 7.
69. Davenport & Montgomery 2000, p. 1.
70. See the proof in Davenport & Montgomery 2000, section 1
71. Iwaniec & Kowalski 2004, p. 1.
72. Varadarajan 2006, sections 2.5, 3.1 and 6.1.
73. Granville 2008, pp. 322–348.
74. See the comment on the importance of modularity in Iwaniec & Kowalski 2004, p. 1
75. Goldfeld 2003.
76. See, e.g., the initial comment in Iwaniec & Kowalski 2004, p. 1.
77. Granville 2008, section 1: "The main difference is that in algebraic number theory [...] one typically considers questions with answers that are given by exact formulas, whereas in analytic number theory [...] one looks for good approximations."
78. See the remarks in the introduction to Iwaniec & Kowalski 2004, p. 1: "However much stronger...".
79. Granville 2008, section 3: "[Riemann] defined what we now call the Riemann zeta function [...] Riemann's deep work gave birth to our subject [...]"
80. See, e.g., Montgomery & Vaughan 2007, p. 1.
81. CITEREFMilne2014, p. 2.
82. Edwards 2000, p. 79.
83. Davis, Martin; Matiyasevich, Yuri; Robinson, Julia (1976). "Hilbert's Tenth Problem: Diophantine Equations: Positive Aspects of a Negative Solution". In Felix E. Browder. Mathematical Developments Arising from Hilbert Problems.

Proceedings of Symposia in Pure Mathematics. XXVIII.2. American Mathematical Society. pp. 323–378. ISBN 0-8218-1428-1. Zbl 0346.02026. Reprinted in The Collected Works of Julia Robinson, Solomon Feferman, editor, pp.269–378, American Mathematical Society 1996.

84. "The Unreasonable Effectiveness of Number Theory", Stefan Andrus Burr, George E. Andrews, American Mathematical Soc., 1992, ISBN 978-0-8218-5501-0
85. Computer science and its relation to mathematics" DE Knuth – The American Mathematical Monthly, 1974
86. "Applications of number theory to numerical analysis", Lo-keng Hua, Luogeng Hua, Yuan Wang, Springer-Verlag, 1981, ISBN 978-3-540-10382-0
87. "Practical applications of algebraic number theory". Mathoverflow.net. Retrieved 2012-05-18.
88. "Where is number theory used in the rest of mathematics?". Mathoverflow.net. 2008-09-23. Retrieved 2012-05-18.
89. Apostol n.d.

Geometry Concepts

Chapter Contents

Geometry is perhaps the most elementary of the sciences that enable man, by purely intellectual processes, to make predictions (based on observation) about physical world. The power of geometry, in the sense of accuracy and utility of these deductions, is impressive, and has been a powerful motivation for the study of logic in geometry.

INTRODUCTION

Geometry (from the Ancient Greek: γεωμετρία; geo- "earth", -metron "measurement") is a branch of mathematics concerned with questions of shape, size, relative position of figures, and the properties of space. A mathematician who works in the field of geometry is called a geometer.

Geometry arose independently in a number of early cultures as a practical way for dealing with lengths, areas, and volumes. Geometry began to see elements of formal mathematical science emerging in the West as early as the 6th century BC. By the 3rd century BC, geometry was put into an axiomatic form by Euclid, whose treatment, Euclid's Elements, set a standard for many centuries to follow. Geometry arose independently in India, with texts providing rules for geometric constructions appearing as early as the 3rd century BC. Islamic scientists preserved Greek ideas and expanded on them during the Middle Ages. By the early 17th century, geometry had been put on a solid analytic footing by mathematicians such as René Descartes and Pierre de Fermat. Since then, and into modern times, geometry has expanded into non-Euclidean geometry and manifolds, describing spaces that lie beyond the normal range of human experience.

While geometry has evolved significantly throughout the years, there are some general concepts that are more or less fundamental to geometry. These include the concepts of points, lines, planes, surfaces, angles, and curves, as well as the more advanced notions of manifolds and topology or metric.

Contemporary geometry has many subfields:

- Euclidean geometry is geometry in its classical sense. The mandatory educational curriculum of the majority of nations includes the study of points, lines, planes, angles, triangles, congruence, similarity, solid figures, circles, and analytic geometry. Euclidean geometry also has applications in computer science, crystallography, and various branches of modern mathematics.
- Differential geometry uses techniques of calculus and linear algebra to study problems in geometry. It has applications in physics, including in general relativity.
- Topology is the field concerned with the properties of geometric objects that are unchanged by continuous mappings. In practice, this often means dealing with large-scale properties of spaces, such as connectedness and compactness.
- Convex geometry investigates convex shapes in the Euclidean space and its more abstract analogues, often using techniques of real analysis. It has close connections to convex analysis, optimization and functional analysis and important applications in number theory.
- Algebraic geometry studies geometry through the use of multivariate polynomials and other algebraic techniques. It has applications in many areas, including cryptography and string theory.
- Discrete geometry is concerned mainly with questions of relative position of simple geometric objects, such as points, lines and circles. It shares many methods and principles with combinatorics.

Geometry has applications to many fields, including art, architecture, physics, as well as to other branches of mathematics.

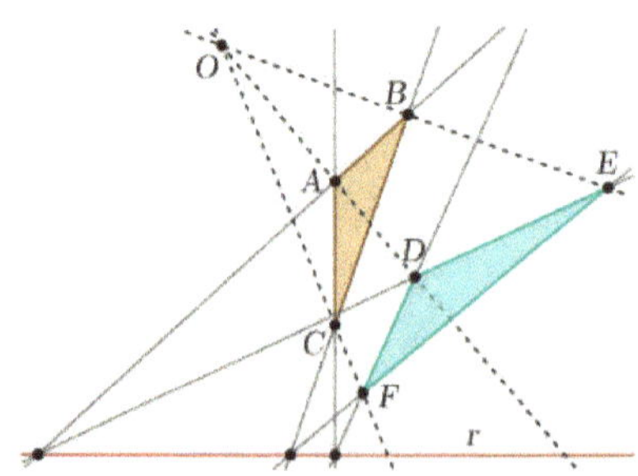

An illustration of Desargues' theorem, an important result in Euclidean and projective geometry

The earliest recorded beginnings of geometry can be traced to ancient Mesopotamia and Egypt in the 2nd millennium BC. Early geometry was a collection of empirically discovered principles concerning lengths, angles, areas, and volumes, which were developed to meet some practical need in surveying, construction, astronomy, and various crafts. The earliest known texts on geometry are the Egyptian Rhind Papyrus (2000–1800 BC) and Moscow Papyrus (c. 1890 BC), the Babylonian clay tablets such as Plimpton 322 (1900 BC). For example, the Moscow Papyrus gives a formula for calculating the volume of a truncated pyramid, or frustum. Later clay tablets (350–50 BC) demonstrate that Babylonian astronomers implemented trapezoid procedures for computing Jupiter's position and motion within time-velocity space. These geometric procedures anticipated the Oxford Calculators, including the mean speed theorem, by 14 centuries. South of Egypt the ancient Nubians established a system of geometry including early versions of sun clocks.

In the 7th century BC, the Greek mathematician Thales of Miletus used geometry to solve problems such as calculating the height of pyramids and the distance of ships from the shore. He is credited with the first use of deductive reasoning applied to geometry, by deriving four corollaries to Thales' Theorem. Pythagoras established the Pythagorean School, which is credited with the first proof of the Pythagorean theorem, though the statement of the theorem has a long history Eudoxus (408–c. 355 BC) developed the method of exhaustion, which allowed the calculation of areas and volumes of curvilinear figures, as well as a theory of ratios that avoided the problem of incommensurable magnitudes, which enabled subsequent geometers to make significant advances. Around 300 BC, geometry was revolutionized by Euclid, whose Elements, widely considered the most successful and influential textbook of all time, introduced mathematical rigor through the axiomatic method and is the earliest example of the format still used in mathematics today, that of definition, axiom, theorem, and proof. Although most of the contents of the Elements were already known, Euclid arranged them into a single, coherent logical framework. The Elements was known to all educated people in the West until the middle of the 20th century and its contents are still taught in geometry classes today. Archimedes (c. 287–212 BC) of Syracuse used the method of exhaustion to calculate the area under the arc of a parabola with the summation of an infinite series, and gave remarkably accurate approximations of Pi. He also studied the spiral bearing his name and obtained formulas for the volumes of surfaces of revolution.

A European and an Arab practicing geometry in the 15th century.

Indian mathematicians also made many important contributions in geometry. The Satapatha Brahmana (3rd century BC) contains rules for ritual geometric constructions that are similar to the Sulba Sutras. According to (Hayashi 2005, p. 363), the Śulba Sūtras contain "the earliest extant verbal expression of the Pythagorean Theorem in the world, although it had already been known to the Old Babylonians. They contain lists of Pythagorean triples, which are particular cases of Diophantine equations.In the Bakhshali manuscript, there is a handful of geometric problems (including problems about volumes of irregular solids). The Bakhshali manuscript also "employs a decimal place value system with a dot for zero." Aryabhata's Aryabhatiya (499) includes the computation of areas and volumes. Brahmagupta wrote his astronomical work Brāhma Sphua Siddhānta in 628. Chapter 12, containing 66 Sanskrit verses, was divided into two sections: "basic operations" (including cube roots, fractions, ratio and proportion, and barter) and "practical mathematics" (including mixture, mathematical series, plane figures, stacking bricks, sawing of timber, and piling of grain). In the latter section, he stated his famous theorem on the diagonals of a cyclic quadrilateral. Chapter 12 also included a formula for the area of a cyclic quadrilateral (a generalization of Heron's formula), as well as a complete description of rational triangles (i.e. triangles with rational sides and rational areas).

In the Middle Ages, mathematics in medieval Islam contributed to the development of geometry, especially algebraic geometry. Al-Mahani (b. 853) conceived the idea of reducing geometrical problems such as duplicating the cube to problems in algebra. Thābit ibn Qurra (known as Thebit in Latin) (836–901) dealt with arithmetic operations applied to ratios of geometrical quantities, and contributed to the development of analytic geometry. Omar Khayyám (1048–1131) found geometric solutions to cubic equations. The theorems of Ibn al-Haytham (Alhazen), Omar Khayyam and Nasir al-Din al-Tusi on quadrilaterals, including the Lambert quadrilateral and Saccheri quadrilateral, were early results in hyperbolic geometry, and along with their alternative postulates, such as Playfair's axiom, these works had a considerable influence on the development of non-Euclidean geometry among later European geometers, including Witelo (c. 1230–c. 1314), Gersonides (1288–1344), Alfonso, John Wallis, and Giovanni Girolamo Saccheri.

In the early 17th century, there were two important developments in geometry. The first was the creation of analytic geometry, or geometry with coordinates and equations, by René Descartes (1596–1650) and Pierre de Fermat (1601–1665). This was a necessary precursor to the development of calculus and a precise quantitative science of physics. The second geometric development of this period was the systematic study of projective geometry by Girard Desargues (1591–1661). Projective geometry is a geometry without measurement or parallel lines, just the study of how points are related to each other.

Two developments in geometry in the 19th century changed the way it had been studied previously. These were the discovery of non-Euclidean geometries by Nikolai Ivanovich Lobachevsky, János Bolyai and Carl Friedrich Gauss and of the formulation of symmetry as the central consideration in the Erlangen Programme of Felix Klein (which generalized the Euclidean and non-Euclidean geometries). Two of the master geometers of the time were Bernhard Riemann (1826–1866), working primarily with tools from mathematical analysis, and introducing the Riemann surface, and Henri Poincaré, the founder of algebraic topology and the geometric theory of dynamical systems. As a consequence of these major changes in the conception of geometry, the concept of "space" became something rich and varied, and the natural background for theories as different as complex analysis and classical mechanics.

Woman teaching geometry. Illustration at the beginning of a medieval translation of Euclid's Elements, (c. 1310)

IMPORTANT CONCEPTS IN GEOMETRY

The following are some of the most important concepts in geometry.

Axioms

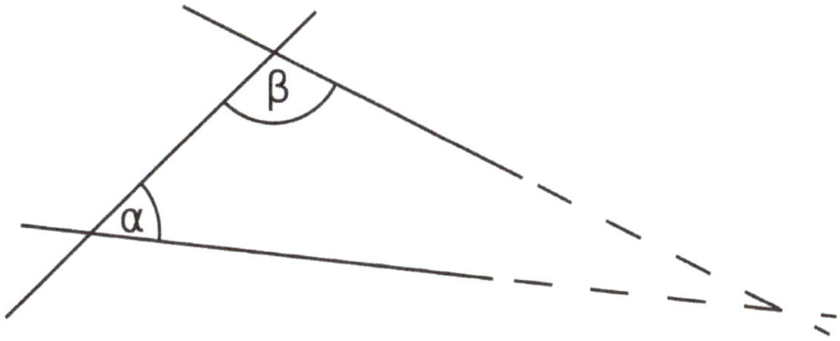

An illustration of Euclid's parallel postulate

Euclid took an abstract approach to geometry in his Elements, one of the most influential books ever written. Euclid introduced certain axioms, or postulates, expressing primary or self-evident properties of points, lines, and planes. He proceeded to rigorously deduce other properties by mathematical reasoning. The characteristic feature of Euclid's approach to geometry was its rigor, and it has come to be known as axiomatic or synthetic geometry. At the start of the 19th century, the discovery of non-Euclidean geometries by Nikolai Ivanovich Lobachevsky (1792–1856), János Bolyai (1802–1860), Carl Friedrich Gauss (1777–1855) and others led to a revival of interest in this discipline, and in the 20th century, David Hilbert (1862–1943) employed axiomatic reasoning in an attempt to provide a modern foundation of geometry.

Points

Points are considered fundamental objects in Euclidean geometry. They have been defined in a variety of ways, including Euclid's definition as 'that which has no part' and through the use of algebra or nested sets. In many areas of geometry, such as analytic geometry, differential geometry, and topology, all objects are considered to be built up from points. However, there has been some study of geometry without reference to points.

Lines

Euclid described a line as "breadthless length" which "lies equally with respect to the points on itself". In modern mathematics, given the multitude of geometries, the concept of a line is closely tied to the way the geometry is described. For instance, in analytic geometry, a line in the plane is often defined as the set of points whose coordinates satisfy a given linear equation, but in a more abstract setting, such as incidence geometry, a line may be an independent object, distinct from the set of points which lie on it. In differential geometry, a geodesic is a generalization of the notion of a line to curved spaces.

Planes

A plane is a flat, two-dimensional surface that extends infinitely far. Planes are used in every area of geometry. For instance, planes can be studied as a topological surface without reference to distances or angles; it can be studied as an affine space, where collinearity and ratios can be studied but not distances; it can be studied as the complex plane using techniques of complex analysis; and so on.

Angles

Euclid defines a plane angle as the inclination to each other, in a plane, of two lines which meet each other, and do not lie straight with respect to each other. In modern terms, an angle is the figure formed by two rays, called the sides of the angle, sharing a common endpoint, called the vertex of the angle.

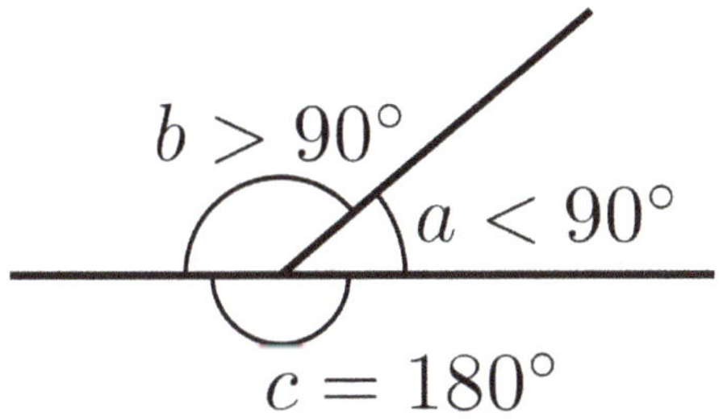

Acute (a), obtuse (b), and straight (c) angles. The acute and obtuse angles are also known as oblique angles.

In Euclidean geometry, angles are used to study polygons and triangles, as well as forming an object of study in their own right. The study of the angles of a triangle or of angles in a unit circle forms the basis of trigonometry.

In differential geometry and calculus, the angles between plane curves or space curves or surfaces can be calculated using the derivative.

Curves

A curve is a 1-dimensional object that may be straight (like a line) or not; curves in 2-dimensional space are called plane curves and those in 3-dimensional space are called space curves.

In topology, a curve is defined by a function from an interval of the real numbers to another space. In differential geometry, the same definition is used, but the defining function is required to be differentiable Algebraic geometry studies algebraic curves, which are defined as algebraic varieties of dimension one.

Surfaces

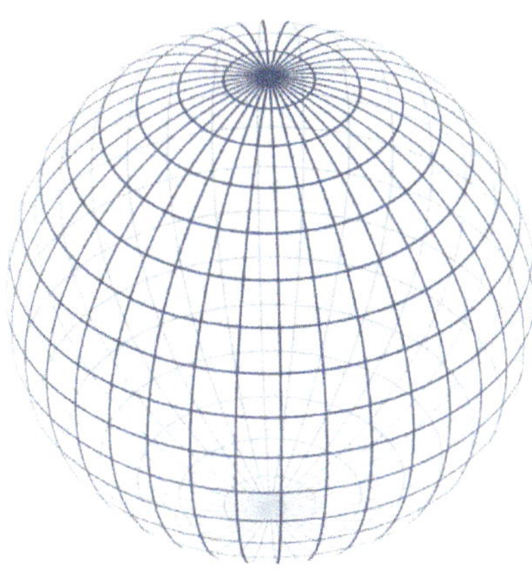

A sphere is a surface that can be defined parametrically (by $x = r \sin \theta \cos \varphi$, $y = r \sin \theta \sin \varphi$, $z = r \cos \theta$) or implicitly (by $x^2 + y^2 + z^2 - r^2 = 0$.)

A surface is a two-dimensional object, such as a sphere or paraboloid. In differential geometry and topology, surfaces are described by two-dimensional 'patches' (or neighborhoods) that are assembled by diffeomorphisms or homeomorphisms, respectively. In algebraic geometry, surfaces are described by polynomial equations.

Manifolds

A manifold is a generalization of the concepts of curve and surface. In topology, a manifold is a topological space where every point has a neighborhood that is homeomorphic to Euclidean space. In differential geometry, a differentiable manifold is a space where each neighborhood is diffeomorphic to Euclidean space.

Manifolds are used extensively in physics, including in general relativity and string theory

Topologies and Metrics

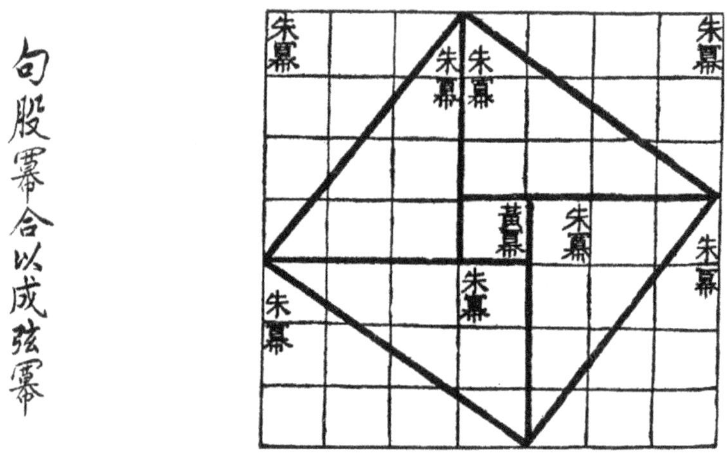

Visual checking of the Pythagorean theorem for the (3, 4, 5) triangle as in the Chou Pei Suan Ching 500–200 BC. The Pythagorean theorem is a consequence of the Euclidean metric.

A topology is a mathematical structure on a set that tells how elements of the set relate spatially to each other. The best-known examples of topologies come from metrics, which are ways of measuring distances between points. For instance, the Euclidean metric measures the distance between points in the Euclidean plane, while the hyperbolic metric measures the distance in the hyperbolic plane. Other important examples of metrics include the Lorentz metric of special relativity and the semi-Riemannian metrics of general relativity.

Compass and Straightedge Constructions

Classical geometers paid special attention to constructing geometric objects that had been described in some other way. Classically, the only instruments allowed in geometric constructions are the compass and straightedge. Also, every construction had to be complete in a finite number of steps. However, some problems turned out to be difficult or impossible to solve by these means alone, and ingenious constructions using parabolas and other curves, as well as mechanical devices, were found.

Dimension

The Koch snowflake, with fractal dimension=log4/log3 and topological dimension=1

Where the traditional geometry allowed dimensions 1 (a line), 2 (a plane) and 3 (our ambient world conceived of as three-dimensional space), mathematicians have used higher dimensions for nearly two centuries. Dimension has gone through stages of being any natural number n, possibly infinite with the introduction of Hilbert space, and any positive real number in fractal geometry. Dimension theory is a technical area, initially within general topology, that discusses definitions; in common with most mathematical ideas, dimension is now defined rather than an intuition. Connected topological manifolds have a well-defined dimension; this is a theorem (invariance of domain) rather than anything a priori.

The issue of dimension still matters to geometry, in the absence of complete answers to classic questions. Dimensions 3 of space and 4 of space-time are special cases in geometric topology. Dimension 10 or 11 is a key number in string theory. Research may bring a satisfactory geometric reason for the significance of 10 and 11 dimensions.

Symmetry

A tiling of the hyperbolic plane

The theme of symmetry in geometry is nearly as old as the science of geometry itself. Symmetric shapes such as the circle, regular polygons and platonic solids held deep significance for many ancient philosophers and were investigated in detail before the time of Euclid. Symmetric patterns occur in nature and were artistically rendered in a multitude of forms, including the graphics of M. C. Escher. Nonetheless, it was not until the second half of 19th century that the unifying role of symmetry in foundations of geometry was recognized. Felix Klein's Erlangen program proclaimed that, in a very precise sense, symmetry, expressed via the notion of a transformation group, determines what geometry is. Symmetry in classical Euclidean geometry is represented by congruences and rigid motions, whereas in projective geometry an analogous role is played by collineations, geometric transformations that take straight lines into straight lines. However it was in the new geometries of Bolyai and Lobachevsky, Riemann, Clifford and Klein, and Sophus Lie that Klein's idea to 'define a geometry via its symmetry group' proved most influential. Both discrete and continuous symmetries play prominent roles in geometry, the former in topology and geometric group theory, the latter in Lie theory and Riemannian geometry.

A different type of symmetry is the principle of duality in projective geometry (see Duality (projective geometry)) among other fields. This meta-phenomenon can roughly be described as follows: in any theorem, exchange point with plane, join with meet, lies in with contains, and you will get an equally true theorem. A similar and closely related form of duality exists between a vector space and its dual space.

Non-Euclidean Geometry

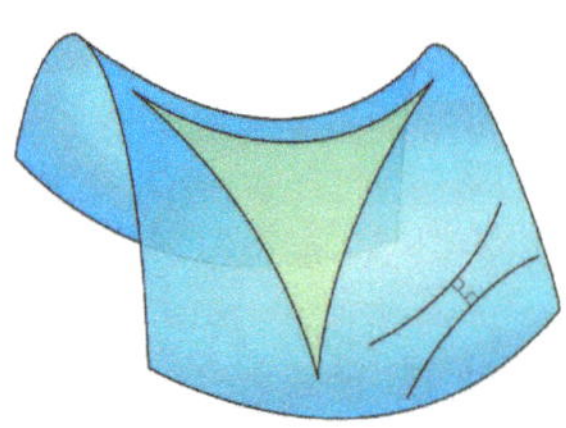

Differential geometry uses tools from calculus to study problems involving curvature.

In the nearly two thousand years since Euclid, while the range of geometrical questions asked and answered inevitably expanded, the basic understanding of space remained essentially the same. Immanuel Kant argued that there is only one, absolute, geometry, which is known to be true a priori by an inner faculty of mind: Euclidean geometry was synthetic a priori. This dominant view was overturned by the revolutionary discovery of non-Euclidean geometry in the works of Bolyai, Lobachevsky, and Gauss (who never published his theory). They demonstrated that ordinary Euclidean space is only one possibility for development of geometry. A broad vision of the subject of geometry was then expressed by Riemann in his 1867 inauguration lecture Über die Hypothesen, welche der Geometrie zu Grunde liegen (On the hypotheses on which geometry is based), published only after his death. Riemann's new idea of space proved crucial in Einstein's general relativity theory, and Riemannian geometry, that considers very general spaces in which the notion of length is defined, is a mainstay of modern geometry.

CONTEMPORARY GEOMETRY

Euclidean Geometry

Geometry lessons in the 20th century

Euclidean geometry has become closely connected with computational geometry, computer graphics, convex geometry, incidence geometry, finite geometry, discrete geometry, and some areas of combinatorics. Attention was given to further work on Euclidean geometry and the Euclidean groups by crystallography and the work of H. S. M. Coxeter, and can be seen in theories of Coxeter groups and polytopes. Geometric group theory is an expanding area of the theory of more general discrete groups, drawing on geometric models and algebraic techniques.

Differential Geometry

Differential geometry has been of increasing importance to mathematical physics due to Einstein's general relativity postulation that the universe is curved. Contemporary differential geometry is intrinsic, meaning that the spaces it considers are smooth manifolds whose geometric structure is governed by a Riemannian metric, which determines how distances are measured near each point, and not a priori parts of some ambient flat Euclidean space.

Topology and Geometry

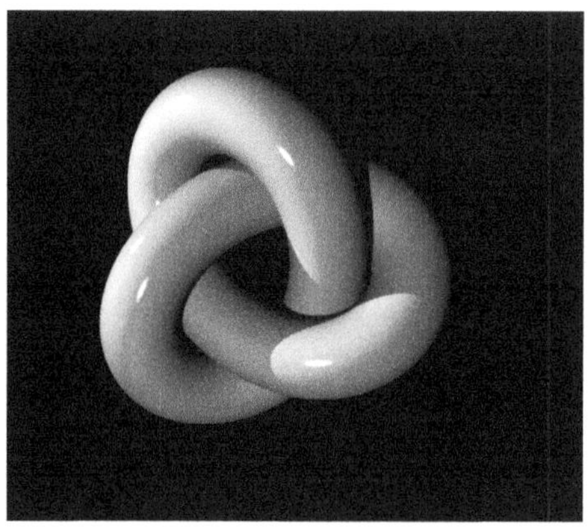

A thickening of the trefoil knot

The field of topology, which saw massive development in the 20th century, is in a technical sense a type of transformation geometry, in which transformations are homeomorphisms. This has often been expressed in the form of the dictum 'topology is rubber-sheet geometry'. Contemporary geometric topology and differential topology, and particular subfields such as Morse theory, would be counted by most mathematicians as part of geometry. Algebraic topology and general topology have gone their own ways.

Algebraic Geometry

The field of algebraic geometry is the modern incarnation of the Cartesian geometry of co-ordinates. From late 1950s through mid-1970s it had undergone major foundational development, largely due to work of Jean-Pierre Serre and Alexander Grothendieck. This led to the introduction of schemes and greater emphasis on topological methods, including various cohomology theories. One of seven Millennium Prize problems, the Hodge conjecture, is a question in algebraic geometry.

Quintic Calabi–Yau threefold

The study of low-dimensional algebraic varieties, algebraic curves, algebraic surfaces and algebraic varieties of dimension 3 ("algebraic threefolds"), has been far advanced. Gröbner basis theory and real algebraic geometry are among more applied subfields of modern algebraic geometry. Arithmetic geometry is an active field combining algebraic geometry and number theory. Other directions of research involve moduli spaces and complex geometry. Algebro-geometric methods are commonly applied in string and brane theory.

APPLICATIONS

Geometry has found applications in many fields, some of which are described below.

Art

Mathematics and art are related in a variety of ways. For instance, the theory of perspective showed that there is more to geometry than just the metric properties of figures: perspective is the origin of projective geometry.

Architecture

Mathematics and architecture are related, since, as with other arts, architects use mathematics for several reasons. Apart from the mathematics needed when engineering buildings, architects use geometry: to define the spatial form of a building; from the Pythagoreans of the sixth century BC onwards, to create forms considered harmonious, and thus to lay out buildings and their surroundings according to mathematical, aesthetic and sometimes religious principles; to decorate buildings with mathematical objects such as tessellations; and to meet environmental goals, such as to minimise wind speeds around the bases of tall buildings.

Physics

The 4_{21}polytope, orthogonally projected into the E_8Lie group Coxeter plane. Lie groups have several applications in physics.

The field of astronomy, especially as it relates to mapping the positions of stars and planets on the celestial sphere and describing the relationship between movements of celestial bodies, have served as an important source of geometric problems throughout history.

Modern geometry has many ties to physics as is exemplified by the links between pseudo-Riemannian geometry and general relativity. One of the youngest physical theories, string theory, is also very geometric in flavour.

Other Fields of Mathematics

Geometry has also had a large effect on other areas of mathematics. For instance, the introduction of coordinates by René Descartes and the concurrent developments of algebra marked a new stage for geometry, since geometric figures such as plane curves could now be represented analytically in the form of functions and equations. This played a key role in the emergence of infinitesimal calculus in the 17th century. The subject of geometry was further enriched by the study of the intrinsic structure of geometric objects that originated with Euler and Gauss and led to the creation of topology and differential geometry.

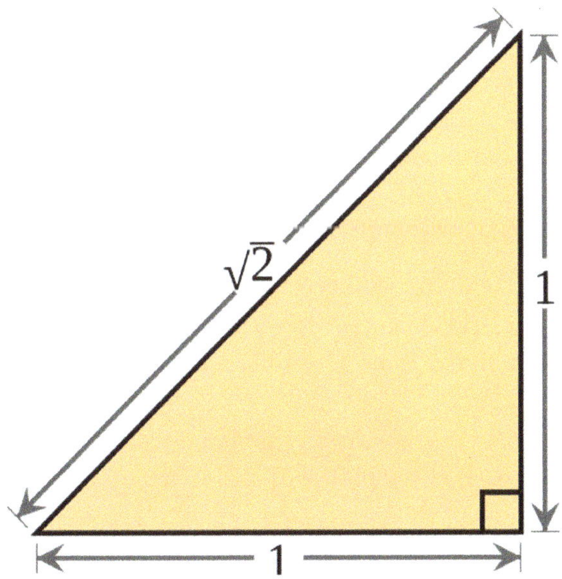

The Pythagoreans discovered that the sides of a triangle could have incommensurable lengths.

An important area of application is number theory. In ancient Greece the Pythagoreans considered the role of numbers in geometry. However, the discovery of incommensurable lengths, which contradicted their philosophical views, made them abandon abstract numbers in favor of concrete geometric quantities, such as length and area of figures. Since the 19th century, geometry has been used for solving problems in number theory, for example through the geometry of numbers or, more recently, scheme theory, which is used in Wiles's proof of Fermat's Last Theorem.

While the visual nature of geometry makes it initially more accessible than other mathematical areas such as algebra or number theory, geometric language is also used in contexts far removed from its traditional, Euclidean provenance (for example, in fractal geometry and algebraic geometry).

Analytic geometry applies methods of algebra to geometric questions, typically by relating geometric curves to algebraic equations. These ideas played a key role in the development of calculus in the 17th century and led to the discovery of many new properties of plane curves. Modern algebraic geometry considers similar questions on a vastly more abstract level.

Leonhard Euler, in studying problems like the Seven Bridges of Königsberg, considered the most fundamental properties of geometric figures based solely on shape, independent of their metric properties. Euler called this new branch of geometry geometria situs (geometry of place), but it is now known as topology. Topology grew out of geometry, but turned into a large independent discipline. It does not differentiate between objects that can be continuously deformed into each other. The objects may nevertheless retain some geometry, as in the case of hyperbolic knots.

REFERENCES

1. (Boyer 1991, "Ionia and the Pythagoreans" p. 43)
2. Martin J. Turner,Jonathan M. Blackledge,Patrick R. Andrews (1998). Fractal geometry in digital imaging. Academic Press. p. 1. ISBN 0-12-703970-8
3. (Staal 1999)[full citation needed]
4. O'Connor, John J.; Robertson, Edmund F., "Al-Sabi Thabit ibn Qurra al-Harrani", MacTutor History of Mathematics archive, University of St Andrews.
5. Lamb, Evelyn (2015-11-08). "By Solving the Mysteries of Shape-Shifting Spaces, Mathematician Wins $3-Million Prize". Scientific American. Retrieved 2016-08-29.
6. Tabak, John (2014). Geometry: the language of space and form. Infobase Publishing. p. xiv. ISBN 081604953X.
7. Schmidt, W., Houang, R., & Cogan, L. (2002). A coherent curriculum. American educator, 26(2), 1-18.
8. J. Friberg, "Methods and traditions of Babylonian mathematics. Plimpton 322, Pythagorean triples, and the Babylonian triangle parameter equations", Historia Mathematica, 8, 1981, pp. 277—318.
9. Neugebauer, Otto (1969) . The Exact Sciences in Antiquity (2 ed.). Dover Publications. ISBN 978-0-486-22332-2. Chap. IV "Egyptian Mathematics and Astronomy", pp. 71–96.
10. (Boyer 1991, "Egypt" p. 19)
11. Ossendrijver, Mathieu (29 Jan 2016). "Ancient Babylonian astronomers calculated Jupiter's position from the area under a time-velocity graph". Science. 351 (6272): 482–484. doi:10.1126/science.aad8085. Retrieved 29 January 2016.
12. The Journal of Egyptian Archaeology. Vol. 84, 1998 Gnomons at Meroë and Early Trigonometry. pg. 171
13. Slayman, Andrew (May 27, 1998). "Neolithic Skywatchers". Archaeology Magazine Archive.
14. Eves, Howard, An Introduction to the History of Mathematics, Saunders, 1990, ISBN 0-03-029558-0.
15. Kurt Von Fritz (1945). "The Discovery of Incommensurability by Hippasus of Metapontum". The Annals of Mathematics.
16. James R. Choike (1980). "The Pentagram and the Discovery of an Irrational Number". The Two-Year College Mathematics Journal.
17. (Boyer 1991, "The Age of Plato and Aristotle" p. 92)
18. (Boyer 1991, "Euclid of Alexandria" p. 119)
19. (Boyer 1991, "Euclid of Alexandria" p. 104)
20. Howard Eves, An Introduction to the History of Mathematics, Saunders, 1990, ISBN 0-03-029558-0 p. 141: "No work, except The Bible, has been more widely used...."

21. O'Connor, J.J.; Robertson, E.F. (February 1996). "A history of calculus". University of St Andrews. Retrieved 2007-08-07.
22. 22. Pythagorean triples are triples of integers with the property: etc.
23. (Cooke 2005, p. 198): "The arithmetic content of the Śulva Sūtras consists of rules for finding Pythagorean triples such as (3, 4, 5), (5, 12, 13), (8, 15, 17), and (12, 35, 37). It is not certain what practical use these arithmetic rules had. The best conjecture is that they were part of religious ritual. A Hindu home was required to have three fires burning at three different altars. The three altars were to be of different shapes, but all three were to have the same area. These conditions led to certain "Diophantine" problems, a particular case of which is the generation of Pythagorean triples, so as to make one square integer equal to the sum of two others."
24. (Hayashi 2005, p. 371)
25. (Hayashi 2003, pp. 121–122)
26. R. Rashed (1994), The development of Arabic mathematics: between arithmetic and algebra, p. 35 London
27. Boyer (1991). "The Arabic Hegemony". A History of Mathematics. pp. 241–242. Omar Khayyam (ca. 1050–1123), the "tent-maker," wrote an Algebra that went beyond that of al-Khwarizmi to include equations of third degree. Like his Arab predecessors, Omar Khayyam provided for quadratic equations both arithmetic and geometric solutions; for general cubic equations, he believed (mistakenly, as the 16th century later showed), arithmetic solutions were impossible; hence he gave only geometric solutions. The scheme of using intersecting conics to solve cubics had been used earlier by Menaechmus, Archimedes, and Alhazan, but Omar Khayyam took the praiseworthy step of generalizing the method to cover all third-degree equations (having positive roots). .. For equations of higher degree than three, Omar Khayyam evidently did not envision similar geometric methods, for space does not contain more than three dimensions, ... One of the most fruitful contributions of Arabic eclecticism was the tendency to close the gap between numerical and geometric algebra. The decisive step in this direction came much later with Descartes, but Omar Khayyam was moving in this direction when he wrote, "Whoever thinks algebra is a trick in obtaining unknowns has thought it in vain. No attention should be paid to the fact that algebra and geometry are different in appearance. Algebras are geometric facts which are proved."
28. O'Connor, John J.; Robertson, Edmund F., "Al-Mahani", MacTutor History of Mathematics archive, University of St Andrews.
29. O'Connor, John J.; Robertson, Edmund F., "Omar Khayyam", MacTutor History of Mathematics archive, University of St Andrews.
30. Boris A. Rosenfeld and Adolf P. Youschkevitch (1996), "Geometry", in Roshdi Rashed, ed., Encyclopedia of the History of Arabic Science, Vol. 2, p. 447–494 , Routledge, London and New York:

31. "Three scientists, Ibn al-Haytham, Khayyam, and al-Tusi, had made the most considerable contribution to this branch of geometry whose importance came to be completely recognized only in the 19th century. In essence, their propositions concerning the properties of quadrangles which they considered, assuming that some of the angles of these figures were acute of obtuse, embodied the first few theorems of the hyperbolic and the elliptic geometries. Their other proposals showed that various geometric statements were equivalent to the Euclidean postulate V. It is extremely important that these scholars established the mutual connection between this postulate and the sum of the angles of a triangle and a quadrangle. By their works on the theory of parallel lines Arab mathematicians directly influenced the relevant investigations of their European counterparts. The first European attempt to prove the postulate on parallel lines – made by Witelo, the Polish scientists of the 13th century, while revising Ibn al-Haytham's Book of Optics (Kitab al-Manazir) – was undoubtedly prompted by Arabic sources. The proofs put forward in the 14th century by the Jewish scholar Levi ben Gerson, who lived in southern France, and by the above-mentioned Alfonso from Spain directly border on Ibn al-Haytham's demonstration. Above, we have demonstrated that Pseudo-Tusi's Exposition of Euclid had stimulated both J. Wallis's and G. Saccheri's studies of the theory of parallel lines."
32. Euclid's Elements – All thirteen books in one volume, Based on Heath's translation, Green Lion Press ISBN 1-888009-18-7.
33. Clark, Bowman L. (Jan 1985). "Individuals and Points". Notre Dame Journal of Formal Logic. 26 (1): 61–75. doi:10.1305/ndjfl/1093870761. Retrieved 29 August 2016.
34. Gerla, G., 1995, "Pointless Geometries" in Buekenhout, F., Kantor, W. eds., Handbook of incidence geometry: buildings and foundations. North-Holland: 1015–31.
35. John Casey (1885) Analytic Geometry of the Point, Line, Circle, and Conic Sections, link from Internet Archive.
36. Buekenhout, Francis (1995), Handbook of Incidence Geometry: Buildings and Foundations, Elsevier B.V.
37. "geodesic – definition of geodesic in English from the Oxford dictionary". OxfordDictionaries.com. Retrieved 2016-01-20.
38. Munkres, James R. Topology. Vol. 2. Upper Saddle River: Prentice Hall, 2000.
39. Szmielew, Wanda. 'From affine to Euclidean geometry: An axiomatic approach.' Springer, 1983.
40. Ahlfors, Lars V. "Complex analysis: an introduction to the theory of analytic functions of one complex variable." 'New York, London' (1953).
41. Sidorov, L.A. (2001), "Angle", in Hazewinkel, Michiel, Encyclopedia of Mathematics, Springer, ISBN 978-1-55608-010-4
42. Gelfand, Izrail Moiseevič, and Mark Saul. "Trigonometry." 'Trigonometry'. Birkhäuser Boston, 2001. 1-20.

43. Stewart, James (2012). Calculus: Early Transcendentals, 7th ed., Brooks Cole Cengage Learning. ISBN 978-0-538-49790-9
44. Jost, Jürgen (2002), Riemannian Geometry and Geometric Analysis, Berlin: Springer-Verlag, ISBN 3-540-42627-2.
45. Baker, Henry Frederick. Principles of geometry. Vol. 2. CUP Archive, 1954.
46. Do Carmo, Manfredo Perdigao, and Manfredo Perdigao Do Carmo. Differential geometry of curves and surfaces. Vol. 2. Englewood Cliffs: Prentice-hall, 1976.
47. Mumford, David (1999). The Red Book of Varieties and Schemes Includes the Michigan Lectures on Curves and Their Jacobians (2nd ed.). Springer-Verlag. ISBN 3-540-63293-X. Zbl 0945.14001.
48. Briggs, William L., and Lyle Cochran Calculus. "Early Transcendentals." 978-0321570567.
49. Yau, Shing-Tung; Nadis, Steve (2010). The Shape of Inner Space: String Theory and the Geometry of the Universe's Hidden Dimensions. Basic Books. ISBN 978-0-465-02023-2.
50. Dmitri Burago, Yu D Burago, Sergei Ivanov, A Course in Metric Geometry, American Mathematical Society, 2001, ISBN 0-8218-2129-6.
51. Wald, Robert M. (1984), General Relativity, University of Chicago Press, ISBN 0-226-87033-2
52. Kline (1972) "Mathematical thought from ancient to modern times", Oxford University Press, p. 1032. Kant did not reject the logical (analytic a priori) possibility of non-Euclidean geometry, see Jeremy Gray, "Ideas of Space Euclidean, Non-Euclidean, and Relativistic", Oxford, 1989; p. 85. Some have implied that, in light of this, Kant had in fact predicted the development of non-Euclidean geometry, cf. Leonard Nelson, "Philosophy and Axiomatics," Socratic Method and Critical Philosophy, Dover, 1965, p. 164.
53. "Ueber die Hypothesen, welche der Geometrie zu Grunde liegen.".
54. It is quite common in algebraic geometry to speak about geometry of algebraic varieties over finite fields, possibly singular. From a naïve perspective, these objects are just finite sets of points, but by invoking powerful geometric imagery and using well developed geometric techniques, it is possible to find structure and establish properties that make them somewhat analogous to the ordinary spheres or cones.

CHAPTER 5

Matrix

Chapter Contents

- **Introduction**
- **Definition**
- **Basic Operations**
- **Linear Equations**
- **Linear Transformations**
- **Square Matrix**
- **Computational Aspects**
- **Abstract Algebraic Aspects And Generalizations**
- **Applications**

A matrix is a concise and useful way of uniquely representing and working with linear transformations. In particular, every linear transformation can be represented by a matrix, and every matrix corresponds to a unique linear transformation. The matrix, and its close relative the determinant, are extremely important concepts in linear algebra,

INTRODUCTION

In mathematics, a matrix (plural matrices) is a rectangular array of numbers, symbols, or expressions, arranged in rows and columns. For example, the dimensions of the matrix below are 2 × 3 (read "two by three"), because there are two rows and three columns.

$$\begin{bmatrix} 1 & 9 & -13 \\ 20 & 5 & -6 \end{bmatrix}$$

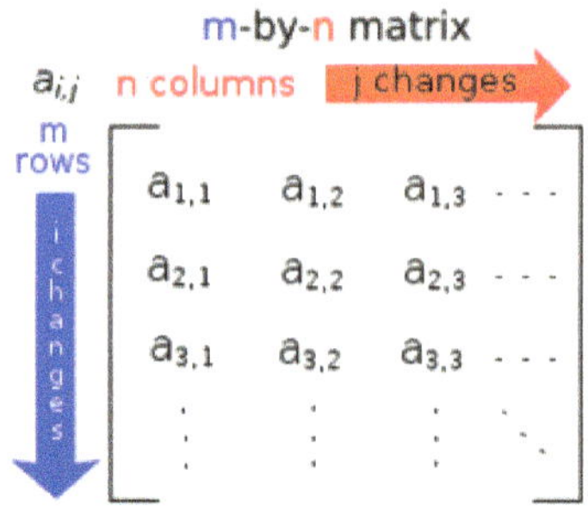

Each element of a matrix is often denoted by a variable with two subscripts. For example, $a_{2,1}$ represents the element at the second row and first column of a matrix **A**.

The individual items in a matrix are called its elements or entries. Provided that they are the same size (have the same number of rows and the same number of columns), two matrices can be added or subtracted element by element. The rule for matrix multiplication, however, is that two matrices can be multiplied only when the number of columns in the first equals the number of rows in the second. Any matrix can be multiplied element-wise by a scalar from its associated field. A major application of matrices is to represent linear transformations, that is, generalizations of linear functions such as $f(x) = 4x$. For example, the rotation of vectors in three dimensional space is a linear transformation which can be represented by a rotation matrix **R**: if **v** is a column vector (a matrix with only one column) describing the position of a point in space, the product **Rv** is a column vector describing the position of that point after a rotation. The product of two transformation matrices is a matrix that represents the composition of two linear transformations. Another application of matrices is in the solution of systems of linear equations. If the matrix is square, it is possible to deduce some of its properties by computing its determinant. For example, a square matrix has an inverse if and only if its determinant is not zero. Insight into the geometry of a linear transformation is obtainable (along with other information) from the matrix's eigenvalues and eigenvectors.

Applications of matrices are found in most scientific fields. In every branch of physics, including classical mechanics, optics, electromagnetism, quantum mechanics, and quantum electrodynamics, they are used to study physical phenomena, such as the motion of rigid bodies. In computer graphics, they are used to project a 3D model onto a 2 dimensional screen. In probability theory and statistics, stochastic matrices are used to describe sets of probabilities; for instance, they are used within the PageRank algorithm that ranks the pages in a Google search. Matrix calculus generalizes classical analytical notions such as derivatives and exponentials to higher dimensions.

A major branch of numerical analysis is devoted to the development of efficient algorithms for matrix computations, a subject that is centuries old and is today an expanding area of research. Matrix decomposition methods simplify computations, both theoretically and practically. Algorithms that are tailored to particular matrix structures, such as sparse matrices and near-diagonal matrices, expedite computations in finite element method and other computations. Infinite matrices occur in planetary theory and in atomic theory. A simple example of an infinite matrix is the matrix representing the derivative operator, which acts on the Taylor series of a function.

DEFINITION

A matrix is a rectangular array of numbers or other mathematical objects for which operations such as addition and multiplication are defined.Most commonly, a matrix over a field F is a rectangular array of scalars each of which is a member of F. Most of this article focuses on real and complex matrices, that is, matrices whose elements are real numbers or complex numbers, respectively. More general types of entries are discussed below. For instance, this is a real matrix:

$$\mathbf{A} = \begin{bmatrix} -1.3 & 0.6 \\ 20.4 & 5.5 \\ 9.7 & -6.2 \end{bmatrix}$$

The numbers, symbols or expressions in the matrix are called its entries or its elements. The horizontal and vertical lines of entries in a matrix are called rows and columns, respectively.

Size

The size of a matrix is defined by the number of rows and columns that it contains. A matrix with m rows and n columns is called an m × n matrix or m-by-n matrix, while m and n are called its dimensions. For example, the matrix **A** above is a 3 × 2 matrix.

Matrices which have a single row are called row vectors, and those which have a single column are called column vectors. A matrix which has the same number of rows and columns is called a square matrix. A matrix with an infinite number of rows or columns (or both) is called an infinite matrix. In some contexts, such as computer algebra programs, it is useful to consider a matrix with no rows or no columns, called an empty matrix.

Name	Size	Example	Description
Row vector	$1 \times n$	$\begin{bmatrix} 3 & 7 & 2 \end{bmatrix}$	A matrix with one row, sometimes used to represent a vector
Column vector	$n \times 1$	$\begin{bmatrix} 4 \\ 1 \\ 8 \end{bmatrix}$	A matrix with one column, sometimes used to represent a vector
Square matrix	$n \times n$	$\begin{bmatrix} 9 & 13 & 5 \\ 1 & 11 & 7 \\ 2 & 6 & 3 \end{bmatrix}$	A matrix with the same number of rows and columns, sometimes used to represent a linear transformation from a vector space to itself, such as reflection, rotation, or shearing.

Notation

Matrices are commonly written in box brackets or parentheses:

$$\mathbf{A} = \begin{bmatrix} a_{11} & a_{12} & \cdots & a_{1n} \\ a_{21} & a_{22} & \cdots & a_{2n} \\ \vdots & \vdots & \ddots & \vdots \\ a_{m1} & a_{m2} & \cdots & a_{mn} \end{bmatrix} = \begin{pmatrix} a_{11} & a_{12} & \cdots & a_{1n} \\ a_{21} & a_{22} & \cdots & a_{2n} \\ \vdots & \vdots & \ddots & \vdots \\ a_{m1} & a_{m2} & \cdots & a_{mn} \end{pmatrix} = (a_{ij}) \in \mathbb{R}^{m \times n}.$$

The specifics of symbolic matrix notation vary widely, with some prevailing trends. Matrices are usually symbolized using upper-case letters (such as **A** in the examples above), while the corresponding lower-case letters, with two subscript indices (for example, a_{11}, or $a_{1,1}$), represent the entries. In addition to using upper-case letters to symbolize matrices, many authors use a special typographical style, commonly boldface upright (non-italic), to further distinguish matrices from other mathematical objects. An alternative notation involves the use of a double-underline with the variable name, with or without boldface style, (for example, $\underline{\underline{A}}$).

The entry in the i-th row and j-th column of a matrix **A** is sometimes referred to as the i,j, (i,j), or (i,j)[th] entry of the matrix, and most commonly denoted as $a_{i,j}$, or a_{ij}. Alternative notations for that entry are A[i,j] or $A_{i,j}$. For example, the (1,3) entry of the following matrix **A** is 5 (also denoted a_{13}, $a_{1,3}$, A[1,3] or $A_{1,3}$):

$$\mathbf{A} = \begin{bmatrix} 4 & -7 & 5 & 0 \\ -2 & 0 & 11 & 8 \\ 19 & 1 & -3 & 12 \end{bmatrix}$$

Sometimes, the entries of a matrix can be defined by a formula such as $a_{i,j} = f(i, j)$. For example, each of the entries of the following matrix **A** is determined by $a_{ij} = i - j$.

$$\mathbf{A} = \begin{bmatrix} 0 & -1 & -2 & -3 \\ 1 & 0 & -1 & -2 \\ 2 & 1 & 0 & -1 \end{bmatrix}$$

In this case, the matrix itself is sometimes defined by that formula, within square brackets or double parentheses. For example, the matrix above is defined as **A** = [i-j], or **A** = ((i-j)). If matrix size is m × n, the above-mentioned formula f(i, j) is valid for any i = 1, ..., m and any j = 1, ..., n. This can be either specified separately, or using m × n as a subscript. For instance, the matrix **A** above is 3 × 4 and can be defined as **A** = [i − j] (i = 1, 2, 3; j = 1, ..., 4), or $\mathbf{A} = [i - j]_{3\times4}$.

Some programming languages utilize doubly subscripted arrays (or arrays of arrays) to represent an m-×-n matrix. Some programming languages start the numbering of array indexes at zero, in which case the entries of an m-by-n matrix are indexed by $0 \le i \le m - 1$ and $0 \le j \le n - 1$. This article follows the more common convention in mathematical writing where enumeration starts from 1.

An asterisk is occasionally used to refer to whole rows or columns in a matrix. For example, $a_{i,*}$ refers to the i[th] row of **A**, and $a_{*,j}$ refers to the j[th]column of **A**. The set of all m-by-n matrices is denoted (m, n).

BASIC OPERATIONS

There are a number of basic operations that can be applied to modify matrices, called matrix addition, scalar multiplication, transposition, matrix multiplication, row operations, and submatrix.

Addition, Scalar Multiplication and Transposition

Operation	Definition	Example
Addition	The *sum* A+B of two *m*-by-*n* matrices A and B is calculated entrywise: $(\mathbf{A}+\mathbf{B})_{i,j} = \mathbf{A}_{i,j} + \mathbf{B}_{i,j}$, where $1 \le i \le m$ and $1 \le j \le n$.	$\begin{bmatrix} 1 & 3 & 1 \\ 1 & 0 & 0 \end{bmatrix} + \begin{bmatrix} 0 & 0 & 5 \\ 7 & 5 & 0 \end{bmatrix} = \begin{bmatrix} 1+0 & 3+0 & 1+5 \\ 1+7 & 0+5 & 0+0 \end{bmatrix} = \begin{bmatrix} 1 & 3 & 6 \\ 8 & 5 & 0 \end{bmatrix}$
Scalar multiplication	The product *c*A of a number *c* (also called a scalar in the parlance of abstract algebra) and a matrix A is computed by multiplying every entry of A by *c*: $(c\mathbf{A})_{i,j} = c \cdot \mathbf{A}_{i,j}$. This operation is called *scalar multiplication*, but its result is not named "scalar product" to avoid confusion, since "scalar product" is sometimes used as a synonym for "inner product".	$2 \cdot \begin{bmatrix} 1 & 8 & -3 \\ 4 & -2 & 5 \end{bmatrix} = \begin{bmatrix} 2\cdot 1 & 2\cdot 8 & 2\cdot -3 \\ 2\cdot 4 & 2\cdot -2 & 2\cdot 5 \end{bmatrix} = \begin{bmatrix} 2 & 16 & -6 \\ 8 & -4 & 10 \end{bmatrix}$
Transposition	The *transpose* of an *m*-by-*n* matrix A is the *n*-by-*m* matrix $\mathbf{A}^T$ (also denoted $\mathbf{A}^{tr}$ or ${}^t\mathbf{A}$) formed by turning rows into columns and vice versa: $(\mathbf{A}^T)_{i,j} = \mathbf{A}_{j,i}$.	$\begin{bmatrix} 1 & 2 & 3 \\ 0 & -6 & 7 \end{bmatrix}^T = \begin{bmatrix} 1 & 0 \\ 2 & -6 \\ 3 & 7 \end{bmatrix}$

Familiar properties of numbers extend to these operations of matrices: for example, addition is commutative, that is, the matrix sum does not depend on the order of the summands: **A** + **B** = **B** + **A**. The transpose is compatible with addition and scalar multiplication, as expressed by $(c\mathbf{A})^T = c(\mathbf{A}^T)$ and $(\mathbf{A} + \mathbf{B})^T = \mathbf{A}^T + \mathbf{B}^T$. Finally, $(\mathbf{A}^T)^T = \mathbf{A}$.

Matrix Multiplication

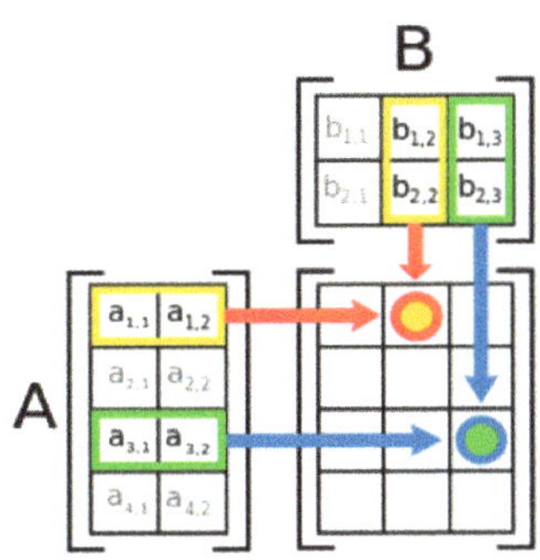

Schematic depiction of the matrix product **AB** of two matrices **A** and **B**.

Multiplication of two matrices is defined if and only if the number of columns of the left matrix is the same as the number of rows of the right matrix. If **A** is an m-by-n matrix and **B** is an n-by-p matrix, then their matrix product**AB** is the m-by-p matrix whose entries are given by dot product of the corresponding row of **A** and the corresponding column of **B**:

$$[\mathbf{AB}]_{i,j} = A_{i,1}B_{1,j} + A_{i,2}B_{2,j} + \cdots + A_{i,n}B_{n,j} = \sum_{r=1}^{n} A_{i,r}B_{r,j}$$

where $1 \leq i \leq m$ and $1 \leq j \leq p$. For example, the underlined entry 2340 in the product is calculated as $(2 \times 1000) + (3 \times 100) + (4 \times 10) = 2340$:

$$\begin{bmatrix} \underline{2} & \underline{3} & \underline{4} \\ 1 & 0 & 0 \end{bmatrix} \begin{bmatrix} 0 & \underline{1000} \\ 1 & \underline{100} \\ 0 & \underline{10} \end{bmatrix} = \begin{bmatrix} 3 & \underline{2340} \\ 0 & 1000 \end{bmatrix}.$$

Matrix multiplication satisfies the rules (AB)C = A(BC) (associativity), and (A+B)C = AC+BC as well as C(A+B) = CA+CB (left and right distributivity), whenever the size of the matrices is such that the various products are defined. The product AB may be defined without BA being defined, namely if A and B are m-by-n and n-by-k matrices, respectively, and $m \neq k$. Even if both products are defined, they need not be equal, that is, generally

AB ≠ BA,

that is, matrix multiplication is not commutative, in marked contrast to (rational, real, or complex) numbers whose product is independent of the order of the factors. An example of two matrices not commuting with each other is:

$$\begin{bmatrix} 1 & 2 \\ 3 & 4 \end{bmatrix} \begin{bmatrix} 0 & 1 \\ 0 & 0 \end{bmatrix} = \begin{bmatrix} 0 & 1 \\ 0 & 3 \end{bmatrix},$$

whereas

$$\begin{bmatrix} 0 & 1 \\ 0 & 0 \end{bmatrix} \begin{bmatrix} 1 & 2 \\ 3 & 4 \end{bmatrix} = \begin{bmatrix} 3 & 4 \\ 0 & 0 \end{bmatrix}.$$

Besides the ordinary matrix multiplication just described, there exist other less frequently used operations on matrices that can be considered forms of multiplication, such as the Hadamard product and the Kronecker product. They arise in solving matrix equations such as the Sylvester equation.

Row Operations

There are three types of row operations:

- row addition, that is adding a row to another.
- row multiplication, that is multiplying all entries of a row by a non-zero constant;
- row switching, that is interchanging two rows of a matrix;

These operations are used in a number of ways, including solving linear equations and finding matrix inverses.

Submatrix

A **submatrix** of a matrix is obtained by deleting any collection of rows and/or columns. For example, from the following 3-by-4 matrix, we can construct a 2-by-3 submatrix by removing row 3 and column 2:

$$\mathbf{A} = \begin{bmatrix} 1 & 2 & 3 & 4 \\ 5 & 6 & 7 & 8 \\ 9 & 10 & 11 & 12 \end{bmatrix} \rightarrow \begin{bmatrix} 1 & 3 & 4 \\ 5 & 7 & 8 \end{bmatrix}.$$

The minors and cofactors of a matrix are found by computing the determinant of certain submatrices.

A **principal submatrix** is a square submatrix obtained by removing certain rows and columns. The definition varies from author to author. According to some authors, a principal submatrix is a submatrix in which the set of row indices that remain is the same as the set of column indices that remain. Other authors define a principal submatrix to be one in which the first k rows and columns, for some number k, are the ones that remain; this type of submatrix has also been called a **leading principal submatrix**.

LINEAR EQUATIONS

Matrices can be used to compactly write and work with multiple linear equations, that is, systems of linear equations. For example, if A is an m-by-n matrix, x designates a column vector (that is, n×1-matrix) of n variables $x_1, x_2, ..., x_n$, and b is an m×1-column vector, then the matrix equation

$\mathbf{Ax} = \mathbf{b}$

is equivalent to the system of linear equations

$A_{1,1}x_1 + A_{1,2}x_2 + ... + A_{1,n}x_n = b_1$

...

$A_{m,1}x_1 + A_{m,2}x_2 + ... + A_{m,n}x_n = b_m$.

LINEAR TRANSFORMATIONS

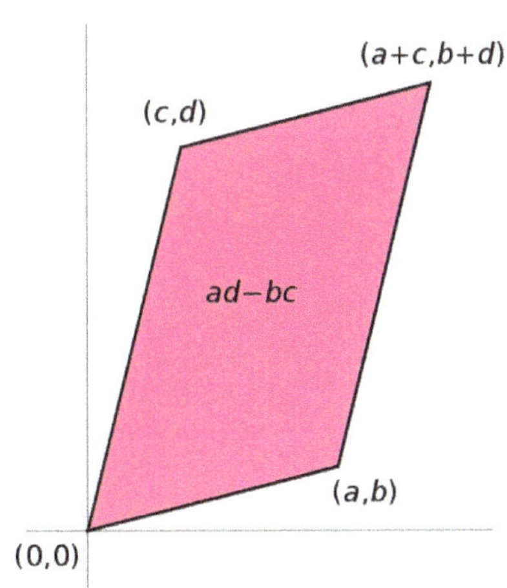

The vectors represented by a 2-by-2 matrix correspond to the sides of a unit square transformed into a parallelogram.

Matrices and matrix multiplication reveal their essential features when related to linear transformations, also known as linear maps. A real m-by-n matrix **A** gives rise to a linear transformation $\mathbf{R}^n \rightarrow \mathbf{R}^m$ mapping each vector **x** in $\mathbf{R}^n$ to the (matrix) product **Ax**, which is a vector in $\mathbf{R}^m$. Conversely, each linear transformation f: $\mathbf{R}^n \rightarrow \mathbf{R}^m$ arises from a unique m-by-n matrix **A**: explicitly, the (i, j)-entry of **A** is the ith coordinate of $f(\mathbf{e}_j)$, where $\mathbf{e}_j$ = (0,...,0,1,0,...,0) is the unit vector with 1 in the jth position and 0 elsewhere. The matrix **A** is said to represent the linear map f, and **A** is called the transformation matrix of f.

For example, the 2×2 matrix

$$\mathbf{A} = \begin{bmatrix} a & c \\ b & d \end{bmatrix}$$

can be viewed as the transform of the unit square into a parallelogram with vertices at (0, 0), (a, b), (a + c, b + d), and (c, d). The parallelogram pictured at the right is obtained by multiplying A with each of the column vectors $\begin{bmatrix} 0 \\ 0 \end{bmatrix}, \begin{bmatrix} 1 \\ 0 \end{bmatrix}, \begin{bmatrix} 1 \\ 1 \end{bmatrix}$ and $\begin{bmatrix} 0 \\ 1 \end{bmatrix}$ in turn. These vectors define the vertices of the unit square.

The following table shows a number of 2-by-2 matrices with the associated linear maps of $\mathbf{R}^2$. The blue original is mapped to the green grid and shapes. The origin (0,0) is marked with a black point.

Horizontal shear with m=1.25.	Reflection through the vertical axis	Squeeze mapping with r=3/2	Scaling by a factor of 3/2	Rotation by $\pi/6^R$ = 30°
$\begin{bmatrix} 1 & 1.25 \\ 0 & 1 \end{bmatrix}$	$\begin{bmatrix} -1 & 0 \\ 0 & 1 \end{bmatrix}$	$\begin{bmatrix} 3/2 & 0 \\ 0 & 2/3 \end{bmatrix}$	$\begin{bmatrix} 3/2 & 0 \\ 0 & 3/2 \end{bmatrix}$	$\begin{bmatrix} \cos(\pi/6^R) & -\sin(\pi/6^R) \\ \sin(\pi/6^R) & \cos(\pi/6^R) \end{bmatrix}$
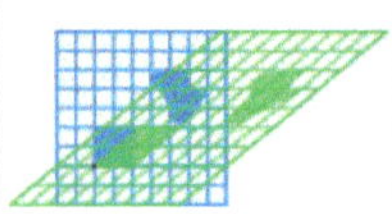	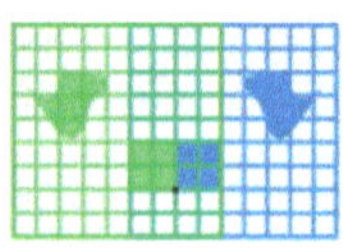	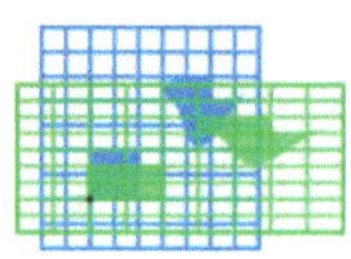	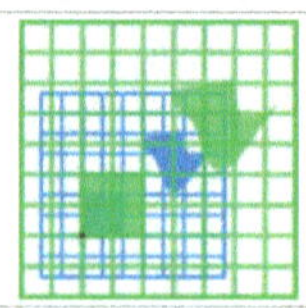	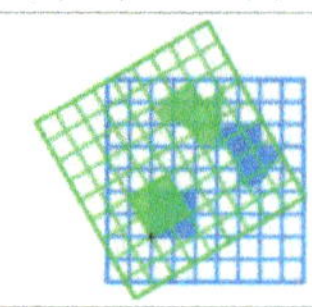

Under the 1-to-1 correspondence between matrices and linear maps, matrix multiplication corresponds to composition of maps: if a k-by-m matrix **B** represents another linear map g : $\mathbf{R}^m \rightarrow \mathbf{R}^k$, then the composition g ∘ f is represented by **BA** since

$(g \circ f)(\mathbf{x}) = g(f(\mathbf{x})) = g(\mathbf{Ax}) = \mathbf{B}(\mathbf{Ax}) = (\mathbf{BA})\mathbf{x}$.

The last equality follows from the above-mentioned associativity of matrix multiplication.

The rank of a matrix **A** is the maximum number of linearly independent row vectors of the matrix, which is the same as the maximum number of linearly independent column vectors. Equivalently it is the dimension of the image of the linear map represented by **A**. The rank-nullity theorem states that the dimension of the kernel of a matrix plus the rank equals the number of columns of the matrix.

SQUARE MATRIX

A square matrix is a matrix with the same number of rows and columns. An n-by-n matrix is known as a square matrix of order n. Any two square matrices of the same order can be added and multiplied. The entries a_{ii} form the main diagonal of a square matrix. They lie on the imaginary line which runs from the top left corner to the bottom right corner of the matrix.

Main Types

Name	Example with $n = 3$
Diagonal matrix	$\begin{bmatrix} a_{11} & 0 & 0 \\ 0 & a_{22} & 0 \\ 0 & 0 & a_{33} \end{bmatrix}$
Lower triangular matrix	$\begin{bmatrix} a_{11} & 0 & 0 \\ a_{21} & a_{22} & 0 \\ a_{31} & a_{32} & a_{33} \end{bmatrix}$
Upper triangular matrix	$\begin{bmatrix} a_{11} & a_{12} & a_{13} \\ 0 & a_{22} & a_{23} \\ 0 & 0 & a_{33} \end{bmatrix}$

Diagonal and Triangular Matrix

If all entries of **A** below the main diagonal are zero, **A** is called an upper triangular matrix. Similarly if all entries of A above the main diagonal are zero, **A** is called a lower triangular matrix. If all off-diagonal elements are zero, **A** is called a diagonal matrix.

Identity Matrix

The identity matrix $\mathbf{I}_n$ of size n is the n-by-n matrix in which all the elements on the main diagonal are equal to 1 and all other elements are equal to 0, for example,

Symmetric or Skew-Symmetric Matrix

A square matrix **A** that is equal to its transpose, that is, $\mathbf{A} = \mathbf{A}^T$, is a symmetric matrix. If instead, **A** is equal to the negative of its transpose, that is, $\mathbf{A} = -\mathbf{A}^T$, then **A** is a skew-symmetric matrix. In complex matrices, symmetry is often replaced by the concept of Hermitian matrices, which satisfy $\mathbf{A}^* = \mathbf{A}$, where the star or asterisk denotes the conjugate transpose of the matrix, that is, the transpose of the complex conjugate of **A**.

By the spectral theorem, real symmetric matrices and complex Hermitian matrices have an eigenbasis; that is, every vector is expressible as a linear combination of eigenvectors. In both cases, all eigenvalues are real. This theorem can be generalized to infinite-dimensional situations related to matrices with infinitely many rows and columns, see below.

Invertible Matrix and its Inverse

A square matrix A is called invertible or non-singular if there exists a matrix B such that

$\mathbf{AB} = \mathbf{BA} = \mathbf{I}_n$.

If **B** exists, it is unique and is called the inverse matrix of **A**, denoted $\mathbf{A}^{-1}$.

Definite Matrix

A symmetric n×n-matrix **A** is called positive-definite (respectively negative-definite; indefinite), if for all nonzero vectors $\mathbf{x} \in \mathbf{R}^n$ the associated quadratic form given by

$Q(x) = x^T A x$

takes only positive values (respectively only negative values; both some negative and some positive values). If the quadratic form takes only non-negative (respectively only non-positive) values, the symmetric matrix is called positive-semidefinite (respectively negative-semidefinite); hence the matrix is indefinite precisely when it is neither positive-semidefinite nor negative-semidefinite.

A symmetric matrix is positive-definite if and only if all its eigenvalues are positive, that is, the matrix is positive-semidefinite and it is invertible. The table at the right shows two possibilities for 2-by-2 matrices.

Allowing as input two different vectors instead yields the bilinear form associated to **A**:

$B_A(\mathbf{x}, \mathbf{y}) = \mathbf{x}^T\mathbf{A}\mathbf{y}$.

Positive definite matrix	Indefinite matrix
$\begin{bmatrix}1/4 & 0\\ 0 & 1\end{bmatrix}$	$\begin{bmatrix}1/4 & 0\\ 0 & -1/4\end{bmatrix}$
$Q(x,y) = 1/4\,x^2 + y^2$	$Q(x,y) = 1/4\,x^2 - 1/4\,y^2$
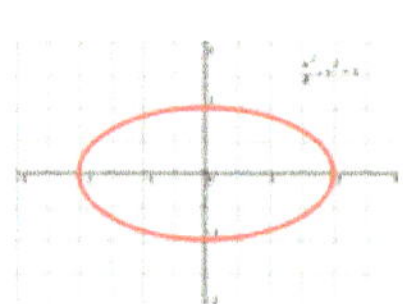 Points such that $Q(x,y)=1$ (Ellipse).	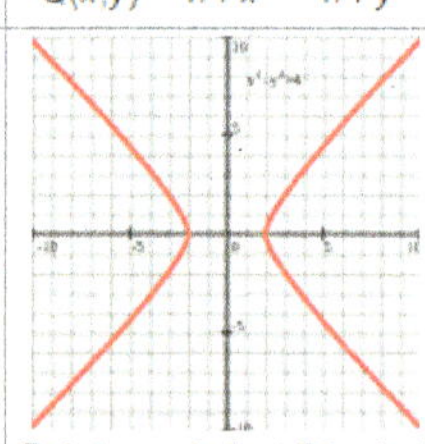 Points such that $Q(x,y)=1$ (Hyperbola).

Orthogonal Matrix

An orthogonal matrix is a square matrix with real entries whose columns and rows are orthogonal unit vectors (that is, orthonormal vectors). Equivalently, a matrix A is orthogonal if its transpose is equal to its inverse:

$$A^{\mathrm{T}} = A^{-1},$$

which entails

$$A^{\mathrm{T}}A = AA^{\mathrm{T}} = I_n,$$

where I is the identity matrix of size n.

An orthogonal matrix A is ncccssarily invcrtiblc (with invcrsc $\Lambda^{-1} - \Lambda^{T}$), unitary ($A^{-1} = A^*$), and normal ($A^*A = AA^*$). The determinant of any orthogonal matrix is either +1 or −1. A special orthogonal matrix is an orthogonal matrix with determinant +1. As a linear transformation, every orthogonal matrix with determinant +1 is a pure rotation, while every orthogonal matrix with determinant -1 is either a pure reflection, or a composition of reflection and rotation.

The complex analogue of an orthogonal matrix is a unitary matrix.

Main Operations

Trace

The trace, tr(**A**) of a square matrix **A** is the sum of its diagonal entries. While matrix multiplication is not commutative as mentioned above, the trace of the product of two matrices is independent of the order of the factors:

tr(**AB**) = tr(**BA**).

This is immediate from the definition of matrix multiplication:

$$\mathrm{tr}(\mathbf{AB}) = \sum_{i=1}^{m}\sum_{j=1}^{n} A_{ij}B_{ji} = \mathrm{tr}(\mathbf{BA}).$$

Also, the trace of a matrix is equal to that of its transpose, that is,

tr(**A**) = tr(**A**^{T}).

Determinant

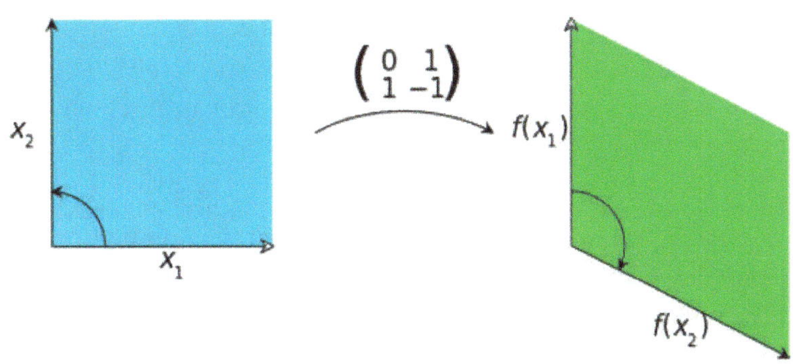

A linear transformation on $\mathbf{R}^2$ given by the indicated matrix. The determinant of this matrix is −1, as the area of the green parallelogram at the right is 1, but the map reverses the orientation, since it turns the counterclockwise orientation of the vectors to a clockwise one.

The determinant det(**A**) or |**A**| of a square matrix **A** is a number encoding certain properties of the matrix. A matrix is invertible if and only if its determinant is nonzero. Its absolute value equals the area (in $\mathbf{R}^2$) or volume (in $\mathbf{R}^3$) of the image of the unit square (or cube), while its sign corresponds to the orientation of the corresponding linear map: the determinant is positive if and only if the orientation is preserved.

The determinant of 2-by-2 matrices is given by

$$\det \begin{bmatrix} a & b \\ c & d \end{bmatrix} = ad - bc.$$

The determinant of 3-by-3 matrices involves 6 terms (rule of Sarrus). The more lengthy Leibniz formula generalises these two formulae to all dimensions.

The determinant of a product of square matrices equals the product of their determinants: $\det(\mathbf{AB}) = \det(\mathbf{A}) \cdot \det(\mathbf{B})$.

Adding a multiple of any row to another row, or a multiple of any column to another column, does not change the determinant. Interchanging two rows or two columns affects the determinant by multiplying it by −1. Using these operations, any matrix can be transformed to a lower (or upper) triangular matrix, and for such matrices the determinant equals the product of the entries on the main diagonal; this provides a method to calculate the determinant of any matrix. Finally, the Laplace expansion expresses the determinant in terms of minors, that is, determinants of smaller matrices. This expansion can be used for a recursive definition of determinants (taking as starting case the determinant of a 1-by-1 matrix, which is its unique entry, or even the determinant of a 0-by-0 matrix, which is 1), that can be seen to be equivalent to the Leibniz formula. Determinants can be used to solve linear systems using Cramer's rule, where the division of the determinants of two related square matrices equates to the value of each of the system's variables.

Eigenvalues and Eigenvectors

A number λ and a non-zero vector **v** satisfying

$$\mathbf{Av} = \lambda\mathbf{v}$$

are called an eigenvalue and an eigenvector of **A**, respectively. The number λ is an eigenvalue of an n×n-matrix **A** if and only if $\mathbf{A}-\lambda\mathbf{I}_n$ is not invertible, which is equivalent to

$$\det(A - \lambda I) = 0.$$

The polynomial p_A in an indeterminate X given by evaluation the determinant $\det(X\mathbf{I}_n-\mathbf{A})$ is called the characteristic polynomial of **A**. It is a monic polynomial of degree n. Therefore the polynomial equation $p_A(\lambda) = 0$ has at most n different solutions, that is, eigenvalues of the matrix. They may be complex even if the entries of **A** are real. According to the Cayley–Hamilton theorem, $p_A(\mathbf{A}) = \mathbf{0}$, that is, the result of substituting the matrix itself into its own characteristic polynomial yields the zero matrix.

COMPUTATIONAL ASPECTS

Matrix calculations can be often performed with different techniques. Many problems can be solved by both direct algorithms or iterative approaches. For example, the eigenvectors of a square matrix can be obtained by finding a sequence of vectors $\mathbf{x}_n$ converging to an eigenvector when n tends to infinity.

To be able to choose the more appropriate algorithm for each specific problem, it is important to determine both the effectiveness and precision of all the available algorithms. The domain studying these matters is called numerical linear algebra. As with other numerical situations, two main aspects are the complexity of algorithms and their numerical stability.

Determining the complexity of an algorithm means finding upper bounds or estimates of how many elementary operations such as additions and multiplications of scalars are necessary to perform some algorithm, for example, multiplication of matrices. For example, calculating the matrix product of two n-by-n matrix using the definition given above needs n^3 multiplications, since for any of the n^2 entries of the product, n multiplications are necessary. The Strassen algorithm outperforms this "naive" algorithm; it needs only $n^{2.807}$ multiplications. A refined approach also incorporates specific features of the computing devices.

In many practical situations additional information about the matrices involved is known. An important case are sparse matrices, that is, matrices most of whose entries are zero. There are specifically adapted algorithms for, say, solving linear systems $\mathbf{Ax} = \mathbf{b}$ for sparse matrices **A**, such as the conjugate gradient method.

An algorithm is, roughly speaking, numerically stable, if little deviations in the input values do not lead to big deviations in the result. For example, calculating the inverse of a matrix via Laplace's formula (Adj (**A**) denotes the adjugate matrix of **A**)

$$\mathbf{A}^{-1} = \mathrm{Adj}(\mathbf{A}) \,/\, \det(\mathbf{A})$$

may lead to significant rounding errors if the determinant of the matrix is very small. The norm of a matrix can be used to capture the conditioning of linear algebraic problems, such as computing a matrix's inverse.

Although most computer languages are not designed with commands or libraries for matrices, as early as the 1970s, some engineering desktop computers such as the HP 9830 had ROM cartridges to add BASIC commands for matrices. Some computer languages such as APL were designed to manipulate matrices, and various mathematical programs can be used to aid computing with matrices.

Decomposition

Main articles: Matrix decomposition, Matrix diagonalization, Gaussian elimination, and Montante's method

There are several methods to render matrices into a more easily accessible form. They are generally referred to as matrix decomposition or matrix factorization techniques. The interest of all these techniques is that they preserve certain properties of the matrices in question, such as determinant, rank or inverse, so that these quantities can be calculated after applying the transformation, or that certain matrix operations are algorithmically easier to carry out for some types of matrices.

The LU decomposition factors matrices as a product of lower (**L**) and an upper triangular matrices (**U**). Once this decomposition is calculated, linear systems can be solved more efficiently, by a simple technique called forward and back substitution. Likewise, inverses of triangular matrices are algorithmically easier to calculate. The Gaussian elimination is a similar algorithm; it transforms any matrix to row echelon form. Both methods proceed by multiplying the matrix by suitable elementary matrices, which correspond to permuting rows or columns and adding multiples of one row to another row. Singular value decomposition expresses any matrix **A** as a product $\mathbf{UDV}^*$, where **U** and **V** are unitary matrices and **D** is a diagonal matrix.

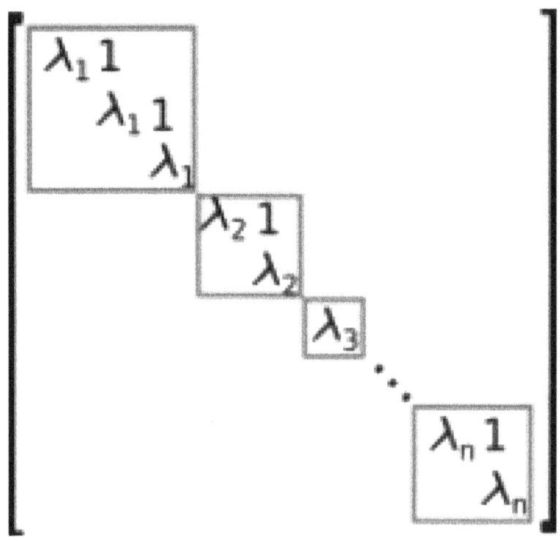

An example of a matrix in Jordan normal form. The grey blocks are called Jordan blocks.

The eigendecomposition or diagonalization expresses **A** as a product $\mathbf{VDV}^{-1}$, where **D** is a diagonal matrix and **V** is a suitable invertible matrix. If **A** can be written in this form, it is called diagonalizable. More generally, and applicable to all matrices, the Jordan decomposition transforms a matrix into Jordan normal form, that is to say matrices whose only nonzero entries are the eigenvalues λ_1 to λ_n of **A**, placed on the main diagonal and possibly entries equal to one directly above the main diagonal, as shown at the right. Given the eigendecomposition, the nth power of **A** (that is, n-fold iterated matrix multiplication) can be calculated via

$$\mathbf{A}^n = (\mathbf{VDV}^{-1})^n = \mathbf{VDV}^{-1}\mathbf{VDV}^{-1}...\mathbf{VDV}^{-1} = \mathbf{VD}^n\mathbf{V}^{-1}$$

and the power of a diagonal matrix can be calculated by taking the corresponding powers of the diagonal entries, which is much easier than doing the exponentiation for **A** instead. This can be used to compute the matrix exponential $e^{\mathbf{A}}$, a need frequently arising in solving linear differential equations, matrix logarithms and square roots of matrices. To avoid numerically ill-conditioned situations, further algorithms such as the Schur decomposition can be employed.

ABSTRACT ALGEBRAIC ASPECTS AND GENERALIZATIONS

Matrices can be generalized in different ways. Abstract algebra uses matrices with entries in more general fields or even rings, while linear algebra codifies properties of matrices in the notion of linear maps. It is possible to consider matrices with infinitely many columns and rows. Another extension are tensors, which can be seen as higher-dimensional arrays of numbers, as opposed to vectors, which can often be realised as sequences of numbers, while matrices are rectangular or two-dimensional arrays of numbers. Matrices, subject to certain requirements tend to form groups known as matrix groups. Similarly under certain conditions matrices form rings known as matrix rings. Though the product of matrices is not in general commutative yet certain matrices form fields known as matrix fields.

Matrices with More General Entries

This article focuses on matrices whose entries are real or complex numbers. However, matrices can be considered with much more general types of entries than real or complex numbers. As a first step of generalization, any field, that is, a set where addition, subtraction, multiplication and division operations are defined and well-behaved, may be used instead of **R** or **C**, for example rational numbers or finite fields. For example, coding theory makes use of matrices over finite fields. Wherever eigenvalues are considered, as these are roots of a polynomial they may exist only in a larger field than that of the entries of the matrix; for instance they may be complex in case of a matrix with real entries. The possibility to reinterpret the entries of a matrix as elements of a larger field (for example, to view a real matrix as a complex matrix whose entries happen to be all real) then allows considering each square matrix to possess a full set of eigenvalues. Alternatively one can consider only matrices with entries in an algebraically closed field, such as **C**, from the outset.

More generally, abstract algebra makes great use of matrices with entries in a ring R. Rings are a more general notion than fields in that a division operation need not exist. The very same addition and multiplication operations of matrices extend to this setting, too. The set M(n, R) of all square n-by-n matrices over R is a ring called matrix ring, isomorphic to the endomorphism ring of the left R-module R^n. If the ring R is commutative, that is, its multiplication is commutative, then M(n, R) is a unitary noncommutative (unless n = 1) associative algebra over R. The determinant of square matrices over a commutative ring R can still be defined using the Leibniz formula; such a matrix is invertible if and only if its determinant is invertible in R, generalising the situation over a field F, where every nonzero element is invertible. Matrices over superrings are called supermatrices.

Matrices do not always have all their entries in the same ring – or even in any ring at all. One special but common case is block matrices, which may be considered as matrices whose entries themselves are matrices. The entries need not be quadratic matrices, and thus need not be members of any ordinary ring; but their sizes must fulfil certain compatibility conditions.

Relationship to Linear Maps

Linear maps $\mathbf{R}^n \to \mathbf{R}^m$ are equivalent to m-by-n matrices, as described above. More generally, any linear map f: V → W between finite-dimensionalvector spaces can be described by a matrix $\mathbf{A} = (a_{ij})$, after choosing bases $\mathbf{v}_1, ..., \mathbf{v}_n$ of V, and $\mathbf{w}_1, ..., \mathbf{w}_m$ of W (so n is the dimension of V and m is the dimension of W), which is such that

$$f(\mathbf{v}_j) = \sum_{i=1}^{m} a_{i,j}\mathbf{w}_i \qquad \text{for } j = 1, \ldots, n.$$

In other words, column j of A expresses the image of $\mathbf{v}_j$ in terms of the basis vectors $\mathbf{w}_i$ of W; thus this relation uniquely determines the entries of the matrix **A**. Note that the matrix depends on the choice of the bases: different choices of bases give rise to different, but equivalent matrices.Many of the above concrete notions can be reinterpreted in this light, for example, the transpose matrix $\mathbf{A}^T$ describes the transpose of the linear map given by **A**, with respect to the dual bases.

These properties can be restated in a more natural way: the category of all matrices with entries in a field k with multiplication as composition is equivalent to the category of finite dimensional vector spaces and linear maps over this field.

More generally, the set of m×n matrices can be used to represent the R-linear maps between the free modules R^m and R^n for an arbitrary ring R with unity. When n = m composition of these maps is possible, and this gives rise to the matrix ring of n×n matrices representing the endomorphism ring of R^n.

Matrix Groups

A group is a mathematical structure consisting of a set of objects together with a binary operation, that is, an operation combining any two objects to a third, subject to certain requirements. A group in which the objects are matrices and the group operation is matrix multiplication is called a matrix group. Since in a group every element has to be invertible, the most general matrix groups are the groups of all invertible matrices of a given size, called the general linear groups.

Any property of matrices that is preserved under matrix products and inverses can be used to define further matrix groups. For example, matrices with a given size and with a determinant of 1 form a subgroup of (that is, a smaller group contained in) their general linear group, called a special linear group. Orthogonal matrices, determined by the condition

$$M^T M = I,$$

form the orthogonal group. Every orthogonal matrix has determinant 1 or −1. Orthogonal matrices with determinant 1 form a subgroup called special orthogonal group.

Every finite group is isomorphic to a matrix group, as one can see by considering the regular representation of the symmetric group. General groups can be studied using matrix groups, which are comparatively well-understood, by means of representation theory.

Infinite Matrices

It is also possible to consider matrices with infinitely many rows and/or columns even if, being infinite objects, one cannot write down such matrices explicitly. All that matters is that for every element in the set indexing rows, and every element in the set indexing columns, there is a well-defined entry (these index sets need not even be subsets of the natural numbers). The basic operations of addition, subtraction, scalar multiplication and transposition can still be defined without problem; however matrix multiplication may involve infinite summations to define the resulting entries, and these are not defined in general.

If R is any ring with unity, then the ring of endomorphisms of $M = \bigoplus_{i \in I} R$ as a right R module is isomorphic to the ring of column finite matrices $\mathbb{CFM}_I(R)$ whose entries are indexed by I × I, and whose columns each contain only finitely many nonzero entries. The endomorphisms of M considered as a left R module result in an analogous object, the row finite matrices $\mathbb{RFM}_I(R)$ whose rows each only have finitely many nonzero entries.

If infinite matrices are used to describe linear maps, then only those matrices can be used all of whose columns have but a finite number of nonzero entries, for the following reason. For a matrix **A** to describe a linear map f: V→W, bases for both spaces must have been chosen; recall that by definition this means that every vector in the space can be written uniquely as a (finite) linear combination of basis vectors, so that written as a (column) vector v of coefficients, only finitely many entries v_i are nonzero. Now the columns of **A** describe the images by f of individual basis vectors of V in the basis of W, which is only meaningful if these columns have only finitely many nonzero entries. There is no restriction on the rows of A however: in the product **A**·v there are only finitely many nonzero coefficients of v involved, so every one of its entries, even if it is given as an infinite sum of products, involves only finitely many nonzero terms and is therefore well defined. Moreover, this amounts to forming a linear combination of the columns of **A** that effectively involves only finitely many of them, whence the result has only finitely many nonzero entries, because each of those columns do. One also sees that products of two matrices of the given type is well defined (provided as usual that the column-index and row-index sets match), is again of the same type, and corresponds to the composition of linear maps.

If R is a normed ring, then the condition of row or column finiteness can be relaxed. With the norm in place, absolutely convergent series can be used instead of finite sums. For example, the matrices whose column sums are absolutely convergent sequences form a ring. Analogously of course, the matrices whose row sums are absolutely convergent series also form a ring.

In that vein, infinite matrices can also be used to describe operators on Hilbert spaces, where convergence and continuity questions arise, which again results in certain constraints that have to be imposed. However, the explicit point of view of matrices tends to obfuscate the matter, and the abstract and more powerful tools of functional analysis can be used instead.

Empty Matrices

An empty matrix is a matrix in which the number of rows or columns (or both) is zero. Empty matrices help dealing with maps involving the zero vector space. For example, if A is a 3-by-0 matrix and B is a 0-by-3 matrix, then AB is the 3-by-3 zero matrix corresponding to the null map from a 3-dimensional space V to itself, while BA is a 0-by-0 matrix. There is no common notation for empty matrices, but most computer algebra systems allow creating and computing with them. The determinant of the 0-by-0 matrix is 1 as follows from regarding the empty product occurring in the Leibniz formula for the determinant as 1. This value is also consistent with the fact that the identity map from any finite dimensional space to itself has determinant 1, a fact that is often used as a part of the characterization of determinants.

APPLICATIONS

There are numerous applications of matrices, both in mathematics and other sciences. Some of them merely take advantage of the compact representation of a set of numbers in a matrix. For example, in game theory and economics, the payoff matrix encodes the payoff for two players, depending on which out of a given (finite) set of alternatives the players choose. Text mining and automated thesaurus compilation makes use of document-term matrices such as tf-idf to track frequencies of certain words in several documents.

Complex numbers can be represented by particular real 2-by-2 matrices via

$$a + ib \leftrightarrow \begin{bmatrix} a & -b \\ b & a \end{bmatrix},$$

under which addition and multiplication of complex numbers and matrices correspond to each other. For example, 2-by-2 rotation matrices represent the multiplication with some complex number of absolute value 1, as above. A similar interpretation is possible for quaternions and Clifford algebras in general.

Early encryption techniques such as the Hill cipher also used matrices. However, due to the linear nature of matrices, these codes are comparatively easy to break. Computer graphics uses matrices both to represent objects and to calculate transformations of objects using affine rotation matrices to accomplish tasks such as projecting a three-dimensional object onto a two-dimensional screen, corresponding to a theoretical camera observation. Matrices over a polynomial ring are important in the study of control theory.

Chemistry makes use of matrices in various ways, particularly since the use of quantum theory to discuss molecular bonding and spectroscopy. Examples are the overlap matrix and the Fock matrix used in solving the Roothaan equations to obtain the molecular orbitals of the Hartree–Fock method.

Graph Theory

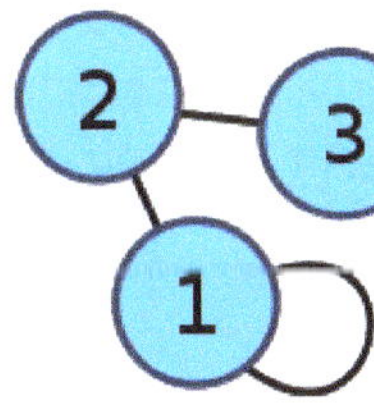

An undirected graph with adjacency matrix

$$\begin{bmatrix} 1 & 1 & 0 \\ 1 & 0 & 1 \\ 0 & 1 & 0 \end{bmatrix}$$

The adjacency matrix of a finite graph is a basic notion of graph theory. It records which vertices of the graph are connected by an edge. Matrices containing just two different values (1 and 0 meaning for example "yes" and "no", respectively) are called logical matrices. The distance (or cost) matrix contains information about distances of the edges. These concepts can be applied to websites connected by hyperlinks or cities connected by roads etc., in which case (unless the connection network is extremely dense) the matrices tend to be sparse, that is, contain few nonzero entries. Therefore, specifically tailored matrix algorithms can be used in network theory.

Analysis and Geometry

The Hessian matrix of a differentiable function f: $\mathbf{R}^n \to \mathbf{R}$ consists of the second derivatives of f with respect to the several coordinate directions, that is,

$$H(f) = \left[\frac{\partial^2 f}{\partial x_i \, \partial x_j} \right]$$

It encodes information about the local growth behaviour of the function: given a critical point $\mathbf{x} = (x_1, ..., x_n)$, that is, a point where the first partial derivatives $\partial f / \partial x_i$ of f vanish, the function has a local minimum if the Hessian matrix is positive definite. Quadratic programming can be used to find global minima or maxima of quadratic functions closely related to the ones attached to matrices (see above).

Another matrix frequently used in geometrical situations is the Jacobi matrix of a differentiable map f: $\mathbf{R}^n \to \mathbf{R}^m$. If $f_1, ..., f_m$ denote the components of f, then the Jacobi matrix is defined as

$$J(f) = \left[\frac{\partial f_i}{\partial x_j} \right]_{1 \le i \le m, 1 \le j \le n}$$

If n > m, and if the rank of the Jacobi matrix attains its maximal value m, f is locally invertible at that point, by the implicit function theorem.

Partial differential equations can be classified by considering the matrix of coefficients of the highest-order differential operators of the equation. For elliptic partial differential equations this matrix is positive definite, which has decisive influence on the set of possible solutions of the equation in question.

The finite element method is an important numerical method to solve partial differential equations, widely applied in simulating complex physical systems. It attempts to approximate the solution to some equation by piecewise linear functions, where the pieces are chosen with respect to a sufficiently fine grid, which in turn can be recast as a matrix equation.

Probability Theory and Statistics

Stochastic matrices are square matrices whose rows are probability vectors, that is, whose entries are non-negative and sum up to one. Stochastic matrices are used to define Markov chains with finitely many states. A row of the stochastic matrix gives the probability distribution for the next position of some particle currently in the state that corresponds to the row. Properties of the Markov chain like absorbing states, that is, states that any particle attains eventually, can be read off the eigenvectors of the transition matrices.

Statistics also makes use of matrices in many different forms. Descriptive statistics is concerned with describing data sets, which can often be represented as data matrices, which may then be subjected to dimensionality reduction techniques. The covariance matrix encodes the mutual variance of several random variables. Another technique using matrices are linear least squares, a method that approximates a finite set of pairs (x_1, y_1), (x_2, y_2), ..., (x_N, y_N), by a linear function

$y_i \approx ax_i + b, i = 1, ..., N$

which can be formulated in terms of matrices, related to the singular value decomposition of matrices.

Random matrices are matrices whose entries are random numbers, subject to suitable probability distributions, such as matrix normal distribution. Beyond probability theory, they are applied in domains ranging from number theory to physics.

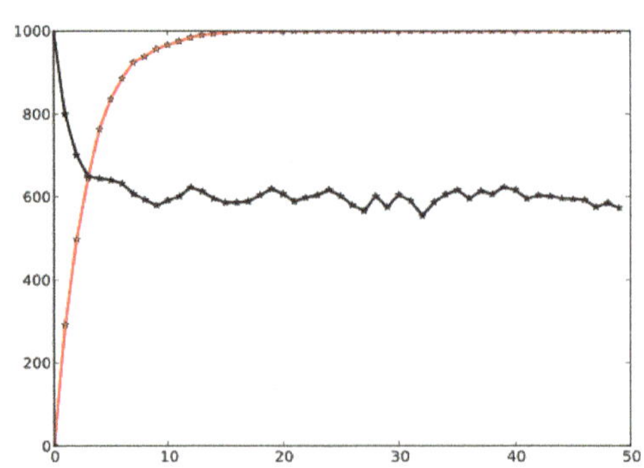

Two different Markov chains. The chart depicts the number of particles (of a total of 1000) in state “2”. Both limiting values can be determined from the transition matrices, which are given by $\begin{bmatrix} .7 & 0 \\ .3 & 1 \end{bmatrix}$ (red) and $\begin{bmatrix} .7 & .2 \\ .3 & .8 \end{bmatrix}$ (black).

Symmetries and Transformations in Physics

Linear transformations and the associated symmetries play a key role in modern physics. For example, elementary particles in quantum field theory are classified as representations of the Lorentz group of special relativity and, more specifically, by their behavior under the spin group. Concrete representations involving the Pauli matrices and more general gamma matrices are an integral part of the physical description of fermions, which behave as spinors. For the three lightest quarks, there is a group-theoretical representation involving the special unitary group SU(3); for their calculations, physicists use a convenient matrix representation known as the Gell-Mann matrices, which are also used for the SU(3) gauge group that forms the basis of the modern description of strong nuclear interactions, quantum chromodynamics. The Cabibbo–Kobayashi–Maskawa matrix, in turn, expresses the fact that the basic quark states that are important for weak interactions are not the same as, but linearly related to the basic quark states that define particles with specific and distinct masses.

Linear Combinations of Quantum States

The first model of quantum mechanics (Heisenberg, 1925) represented the theory’s operators by infinite-dimensional matrices acting on quantum states. This is also referred to as matrix mechanics. One particular example is the density matrix that characterizes the “mixed” state of a quantum system as a linear combination of elementary, “pure” eigenstates.

Another matrix serves as a key tool for describing the scattering experiments that form the cornerstone of experimental particle physics: Collision reactions such as occur in particle accelerators, where non-interacting particles head towards each other and collide in a small interaction zone, with a new set of non-interacting particles as the result, can be described as the scalar product of outgoing particle states and a linear combination of ingoing particle states. The linear combination is given by a matrix known as the S-matrix, which encodes all information about the possible interactions between particles.

Normal Modes

A general application of matrices in physics is to the description of linearly coupled harmonic systems. The equations of motion of such systems can be described in matrix form, with a mass matrix multiplying a generalized velocity to give the kinetic term, and a force matrix multiplying a displacement vector to characterize the interactions. The best way to obtain solutions is to determine the system’s eigenvectors, its normal modes, by diagonalizing the matrix equation. Techniques like this are crucial when it comes to the internal dynamics of molecules: the internal vibrations of systems consisting of mutually bound component atoms. They are also needed for describing mechanical vibrations, and oscillations in electrical circuits.

Geometrical Optics

Geometrical optics provides further matrix applications. In this approximative theory, the wave nature of light is neglected. The result is a model in which light rays are indeed geometrical rays. If the deflection of light rays by optical elements is small, the action of a lens or reflective element on a given light ray can be expressed as multiplication of a two-component vector with a two-by-two matrix called ray transfer matrix: the vector's components are the light ray's slope and its distance from the optical axis, while the matrix encodes the properties of the optical element. Actually, there are two kinds of matrices, viz. a refraction matrix describing the refraction at a lens surface, and a translation matrix, describing the translation of the plane of reference to the next refracting surface, where another refraction matrix applies. The optical system, consisting of a combination of lenses and/or reflective elements, is simply described by the matrix resulting from the product of the components' matrices.

Electronics

Traditional mesh analysis and nodal analysis in electronics lead to a system of linear equations that can be described with a matrix.

The behaviour of many electronic components can be described using matrices. Let A be a 2-dimensional vector with the component's input voltage v_1 and input current i_1 as its elements, and let B be a 2-dimensional vector with the component's output voltage v_2 and output current i_2 as its elements. Then the behaviour of the electronic component can be described by $B = H \cdot A$, where H is a 2 x 2 matrix containing one impedance element (h_{12}), one admittance element (h_{21}) and two dimensionless elements (h_{11} and h_{22}). Calculating a circuit now reduces to multiplying matrices.

Matrices have a long history of application in solving linear equations but they were known as arrays until the 1800s. The Chinese text The Nine Chapters on the Mathematical Art written in 10th–2nd century BCE is the first example of the use of array methods to solve simultaneous equations, including the concept of determinants. In 1545 Italian mathematician Girolamo Cardano brought the method to Europe when he published Ars Magna. The Japanese mathematician Seki used the same array methods to solve simultaneous equations in 1683. The Dutch Mathematician Jan de Witt represented transformations using arrays in his 1659 book Elements of Curves (1659). Between 1700 and 1710 Gottfried Wilhelm Leibniz publicized the use of arrays for recording information or solutions and experimented with over 50 different systems of arrays. Cramer presented his rule in 1750.

The term "matrix" (Latin for "womb", derived from mater—mother) was coined by James Joseph Sylvester in 1850, who understood a matrix as an object giving rise to a number of determinants today called minors, that is to say, determinants of smaller matrices that derive from the original one by removing columns and rows. In an 1851 paper, Sylvester explains:

I have in previous papers defined a "Matrix" as a rectangular array of terms, out of which different systems of determinants may be engendered as from the womb of a common parent.

Arthur Cayley published a treatise on geometric transformations using matrices that were not rotated versions of the coefficients being investigated as had previously been done. Instead he defined operations such as addition, subtraction, multiplication, and division as transformations of those matrices and showed the associative and distributive properties held true. Cayley investigated and demonstrated the non-commutative property of matrix multiplication as well as the commutative property of matrix addition. Early matrix theory had limited the use of arrays almost exclusively to determinants and Arthur Cayley's abstract matrix operations were revolutionary. He was instrumental in proposing a matrix concept independent of equation systems. In 1858 Cayley published his A memoir on the theory of matrices in which he proposed and demonstrated the Cayley-Hamilton theorem.

An English mathematician named Cullis was the first to use modern bracket notation for matrices in 1913 and he simultaneously demonstrated the first significant use of the notation $\mathbf{A} = [a_{i,j}]$ to represent a matrix where $a_{i,j}$ refers to the ith row and the jth column.

The study of determinants sprang from several sources. Number-theoretical problems led Gauss to relate coefficients of quadratic forms, that is, expressions such as $x^2 + xy - 2y^2$, and linear maps in three dimensions to matrices. Eisenstein further developed these notions, including the remark that, in modern parlance, matrix products are non-commutative. Cauchy was the first to prove general statements about determinants, using as definition of the determinant of a matrix $\mathbf{A} = [a_{i,j}]$ the following: replace the powers a_j^k by a_{jk} in the polynomial

$$a_1 a_2 \cdots a_n \prod_{i<j} (a_j - a_i),$$

where Π denotes the product of the indicated terms. He also showed, in 1829, that the eigenvalues of symmetric matrices are real. Jacobi studied "functional determinants"—later called Jacobi determinants by Sylvester—which can be used to describe geometric transformations at a local (or infinitesimal) level, see above; Kronecker's Vorlesungen über die Theorie der Determinanten and Weierstrass' Zur Determinantentheorie, both published in 1903, first treated determinants axiomatically, as opposed to previous more concrete approaches such as the mentioned formula of Cauchy. At that point, determinants were firmly established.

Many theorems were first established for small matrices only, for example the Cayley–Hamilton theorem was proved for 2×2 matrices by Cayley in the aforementioned memoir, and by Hamilton for 4×4 matrices. Frobenius, working on bilinear forms, generalized the theorem to all dimensions (1898). Also at the end of the 19th century the Gauss–Jordan elimination (generalizing a special case now known as Gauss elimination) was established by Jordan. In the early 20th century, matrices attained a central role in linear algebra. partially due to their use in classification of the hypercomplex number systems of the previous century.

The inception of matrix mechanics by Heisenberg, Born and Jordan led to studying matrices with infinitely many rows and columns. Later, von Neumann carried out the mathematical formulation of quantum mechanics, by further developing functional analytic notions such as linear operators on Hilbert spaces, which, very roughly speaking, correspond to Euclidean space, but with an infinity of independent directions.

Other Historical Usages of the Word "Matrix" In Mathematics

The word has been used in unusual ways by at least two authors of historical importance.

Bertrand Russell and Alfred North Whitehead in their Principia Mathematica (1910–1913) use the word "matrix" in the context of their Axiom of reducibility. They proposed this axiom as a means to reduce any function to one of lower type, successively, so that at the "bottom" (0 order) the function is identical to its extension:

"Let us give the name of matrix to any function, of however many variables, which does not involve any apparent variables. Then any possible function other than a matrix is derived from a matrix by means of generalization, that is, by considering the proposition which asserts that the function in question is true with all possible values or with some value of one of the arguments, the other argument or arguments remaining undetermined".

For example, a function $\Phi(x, y)$ of two variables x and y can be reduced to a collection of functions of a single variable, for example, y, by "considering" the function for all possible values of "individuals" a_i substituted in place of variable x. And then the resulting collection of functions of the single variable y, that is, $\forall a_i$: $\Phi(a_i, y)$, can be reduced to a "matrix" of values by "considering" the function for all possible values of "individuals" b_i substituted in place of variable y:

$\forall b_j \forall a_i$: $\Phi(a_i, b_j)$.

Alfred Tarski in his 1946 Introduction to Logic used the word "matrix" synonymously with the notion of truth table as used in mathematical logic.

REFERENCES

1. Anton, Howard (1987), Elementary Linear Algebra (5th ed.), New York: Wiley, ISBN 0-471-84819-0
2. Arnold, Vladimir I.; Cooke, Roger (1992), Ordinary differential equations, Berlin, DE; New York, NY: Springer-Verlag, ISBN 978-3-540-54813-3
3. Artin, Michael (1991), Algebra, Prentice Hall, ISBN 978-0-89871-510-1
4. Association for Computing Machinery (1979), Computer Graphics, Tata McGraw–Hill, ISBN 978-0-07-059376-3
5. Baker, Andrew J. (2003), Matrix Groups: An Introduction to Lie Group Theory, Berlin, DE; New York, NY: Springer-Verlag, ISBN 978-1-85233-470-3
6. Bau III, David; Trefethen, Lloyd N. (1997), Numerical linear algebra, Philadelphia, PA: Society for Industrial and Applied Mathematics, ISBN 978-0-89871-361-9

7. Beauregard, Raymond A.; Fraleigh, John B. (1973), A First Course In Linear Algebra: with Optional Introduction to Groups, Rings, and Fields, Boston: Houghton Mifflin Co., ISBN 0-395-14017-X
8. Bretscher, Otto (2005), Linear Algebra with Applications (3rd ed.), Prentice Hall
9. Bronson, Richard (1970), Matrix Methods: An Introduction, New York: Academic Press, LCCN 70097490
10. Bronson, Richard (1989), Schaum's outline of theory and problems of matrix operations, New York: McGraw–Hill, ISBN 978-0-07-007978-6
11. Brown, William C. (1991), Matrices and vector spaces, New York, NY: Marcel Dekker, ISBN 978-0-8247-8419-5
12. Coburn, Nathaniel (1955), Vector and tensor analysis, New York, NY: Macmillan, OCLC 1029828
13. Conrey, J. Brian (2007), Ranks of elliptic curves and random matrix theory, Cambridge University Press, ISBN 978-0-521-69964-8
14. Fraleigh, John B. (1976), A First Course In Abstract Algebra (2nd ed.), Reading: Addison-Wesley, ISBN 0-201-01984-1
15. Fudenberg, Drew; Tirole, Jean (1983), Game Theory, MIT Press
16. Gilbarg, David; Trudinger, Neil S. (2001), Elliptic partial differential equations of second order (2nd ed.), Berlin, DE; New York, NY: Springer-Verlag, ISBN 978-3-540-41160-4
17. Godsil, Chris; Royle, Gordon (2004), Algebraic Graph Theory, Graduate Texts in Mathematics, 207, Berlin, DE; New York, NY: Springer-Verlag, ISBN 978-0-387-95220-8
18. Golub, Gene H.; Van Loan, Charles F. (1996), Matrix Computations (3rd ed.), Johns Hopkins, ISBN 978-0-8018-5414-9
19. Greub, Werner Hildbert (1975), Linear algebra, Graduate Texts in Mathematics, Berlin, DE; New York, NY: Springer-Verlag, ISBN 978-0-387-90110-7
20. Halmos, Paul Richard (1982), A Hilbert space problem book, Graduate Texts in Mathematics, 19 (2nd ed.), Berlin, DE; New York, NY: Springer-Verlag, ISBN 978-0-387-90685-0, MR 675952
21. Horn, Roger A.; Johnson, Charles R. (1985), Matrix Analysis, Cambridge University Press, ISBN 978-0-521-38632-6
22. Householder, Alston S. (1975), The theory of matrices in numerical analysis, New York, NY: Dover Publications, MR 0378371
23. Kreyszig, Erwin (1972), Advanced Engineering Mathematics (3rd ed.), New York: Wiley, ISBN 0-471-50728-8.
24. Krzanowski, Wojtek J. (1988), Principles of multivariate analysis, Oxford Statistical Science Series, 3, The Clarendon Press Oxford University Press, ISBN 978-0-19-852211-9, MR 969370
25. Itõ, Kiyosi, ed. (1987), Encyclopedic dictionary of mathematics. Vol. I-IV (2nd ed.), MIT Press, ISBN 978-0-262-09026-1, MR 901762

26. Lang, Serge (1969), Analysis II, Addison-Wesley
27. Lang, Serge (1987a), Calculus of several variables (3rd ed.), Berlin, DE; New York, NY: Springer-Verlag, ISBN 978-0-387-96405-8
28. Lang, Serge (1987b), Linear algebra, Berlin, DE; New York, NY: Springer-Verlag, ISBN 978-0-387-96412-6
29. Lang, Serge (2002), Algebra, Graduate Texts in Mathematics, 211 (Revised third ed.), New York: Springer-Verlag, ISBN 978-0-387-95385-4, MR 1878556
30. Latouche, Guy; Ramaswami, Vaidyanathan (1999), Introduction to matrix analytic methods in stochastic modeling (1st ed.), Philadelphia, PA: Society for Industrial and Applied Mathematics, ISBN 978-0-89871-425-8
31. Manning, Christopher D.; Schütze, Hinrich (1999), Foundations of statistical natural language processing, MIT Press, ISBN 978-0-262-13360-9
32. Mehata, K. M.; Srinivasan, S. K. (1978), Stochastic processes, New York, NY: McGraw–Hill, ISBN 978-0-07-096612-3
33. Mirsky, Leonid (1990), An Introduction to Linear Algebra, Courier Dover Publications, ISBN 978-0-486-66434-7
34. Nering, Evar D. (1970), Linear Algebra and Matrix Theory (2nd ed.), New York: Wiley, LCCN 76-91646
35. Nocedal, Jorge; Wright, Stephen J. (2006), Numerical Optimization (2nd ed.), Berlin, DE; New York, NY: Springer-Verlag, p. 449, ISBN 978-0-387-30303-1
36. Oualline, Steve (2003), Practical C++ programming, O'Reilly, ISBN 978-0-596-00419-4
37. Press, William H.; Flannery, Brian P.; Teukolsky, Saul A.; Vetterling, William T. (1992), "LU Decomposition and Its Applications", Numerical Recipes in FORTRAN: The Art of Scientific Computing (PDF) (2nd ed.), Cambridge University Press, pp. 34–42
38. Protter, Murray H.; Morrey, Jr., Charles B. (1970), College Calculus with Analytic Geometry (2nd ed.), Reading: Addison-Wesley, LCCN 76087042
39. Punnen, Abraham P.; Gutin, Gregory (2002), The traveling salesman problem and its variations, Boston, MA: Kluwer Academic Publishers, ISBN 978-1-4020-0664-7
40. Reichl, Linda E. (2004), The transition to chaos: conservative classical systems and quantum manifestations, Berlin, DE; New York, NY: Springer-Verlag, ISBN 978-0-387-98788-0
41. Rowen, Louis Halle (2008), Graduate Algebra: noncommutative view, Providence, RI: American Mathematical Society, ISBN 978-0-8218-4153-2
42. Šolin, Pavel (2005), Partial Differential Equations and the Finite Element Method, Wiley-Interscience, ISBN 978-0-471-76409-0
43. Stinson, Douglas R. (2005), Cryptography, Discrete Mathematics and its Applications, Chapman & Hall/CRC, ISBN 978-1-58488-508-5

44. *Stoer, Josef; Bulirsch, Roland (2002), Introduction to Numerical Analysis (3rd ed.), Berlin, DE; New York, NY: Springer-Verlag, ISBN 978-0-387-95452-3*

45. *Ward, J. P. (1997), Quaternions and Cayley numbers, Mathematics and its Applications,* **403***, Dordrecht, NL: Kluwer Academic Publishers Group, ISBN 978-0-7923-4513-8, MR 1458894*

46. *Wolfram, Stephen (2003), The Mathematica Book (5th ed.), Champaign, IL: Wolfram Media, ISBN 978-1-57955-022-6*

CHAPTER 6

Vector Space

Chapter Contents

- **Introduction**
- **Introduction And Definition**
- **Basic Constructions**
- **Applications**
- **Generalizations**

A vector space V is a set that is closed under finite vector addition and scalar multiplication. The basic example is n-dimensional Euclidean space R^n , where every element is represented by a list of n real numbers, scalars are real numbers, addition is component wise, and scalar multiplication is multiplication on each term separately.

INTRODUCTION

A vector space (also called a linear space) is a collection of objects called vectors, which may be added together and multiplied ("scaled") by numbers, called scalars in this context. Scalars are often taken to be real numbers, but there are also vector spaces with scalar multiplication by complex numbers, rational numbers, or generally any field. The operations of vector addition and scalar multiplication must satisfy certain requirements, called axioms, listed below.

Euclidean vectors are an example of a vector space. They represent physical quantities such as forces: any two forces (of the same type) can be added to yield a third, and the multiplication of a force vector by a real multiplier is another force vector. In the same vein, but in a more geometric sense, vectors representing displacements in the plane or in three-dimensional space also form vector spaces. Vectors in vector spaces do not necessarily have to be arrow-like objects as they appear in the mentioned examples: vectors are regarded as abstract mathematical objects with particular properties, which in some cases can be visualized as arrows.

Vector spaces are the subject of linear algebra and are well characterized by their dimension, which, roughly speaking, specifies the number of independent directions in the space. Infinite-dimensional vector spaces arise naturally in mathematical analysis, as function spaces, whose vectors are functions. These vector spaces are generally endowed with additional structure, which may be a topology, allowing the consideration of issues of proximity and continuity. Among these topologies, those that are defined by a norm or inner product are more commonly used, as having a notion of distance between two vectors. This is particularly the case of Banach spaces and Hilbert spaces, which are fundamental in mathematical analysis.

Historically, the first ideas leading to vector spaces can be traced back as far as the 17th century's analytic geometry, matrices, systems of linear equations, and Euclidean vectors. The modern, more abstract treatment, first formulated by Giuseppe Peano in 1888, encompasses more general objects than Euclidean space, but much of the theory can be seen as an extension of classical geometric ideas like lines, planes and their higher-dimensional analogs.

Today, vector spaces are applied throughout mathematics, science and engineering. They are the appropriate linear-algebraic notion to deal with systems of linear equations; offer a framework for Fourier expansion, which is employed in image compression routines; or provide an environment that can be used for solution techniques for partial differential equations. Furthermore, vector spaces furnish an abstract, coordinate-free way of dealing with geometrical and physical objects such as tensors. This in turn allows the examination of local properties of manifolds by linearization techniques. Vector spaces may be generalized in several ways, leading to more advanced notions in geometry and abstract algebra.

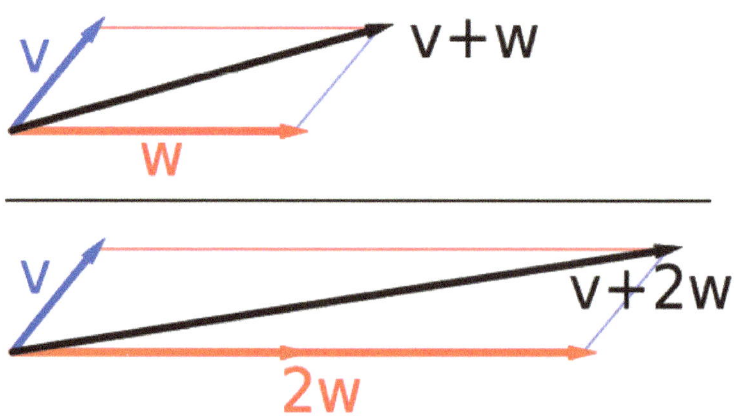

Vector addition and scalar multiplication: a vector **v** (blue) is added to another vector **w** (red, upper illustration). Below, **w** is stretched by a factor of 2, yielding the sum **v** + 2**w**.

INTRODUCTION AND DEFINITION

The concept of vector space will first be explained by describing two particular examples:

First Example: Arrows in the Plane

The first example of a vector space consists of arrows in a fixed plane, starting at one fixed point. This is used in physics to describe forces or velocities. Given any two such arrows, **v** and **w**, the parallelogram spanned by these two arrows contains one diagonal arrow that starts at the origin, too. This new arrow is called the sum of the two arrows and is denoted **v** + **w**. In the special case of two arrows on the same line, their sum is the arrow on this line whose length is the sum or the difference of the lengths, depending on whether the arrows have the same direction. Another operation that can be done with arrows is scaling: given any positive real number a, the arrow that has the same direction as **v**, but is dilated or shrunk by multiplying its length by a, is called multiplication of **v** by a. It is denoted a**v**. When a is negative, a**v** is defined as the arrow pointing in the opposite direction, instead.

The following shows a few examples: if a = 2, the resulting vector a**w** has the same direction as **w**, but is stretched to the double length of **w** (right image below). Equivalently 2**w** is the sum **w** + **w**. Moreover, (−1)**v** = −**v** has the opposite direction and the same length as **v** (blue vector pointing down in the right image).

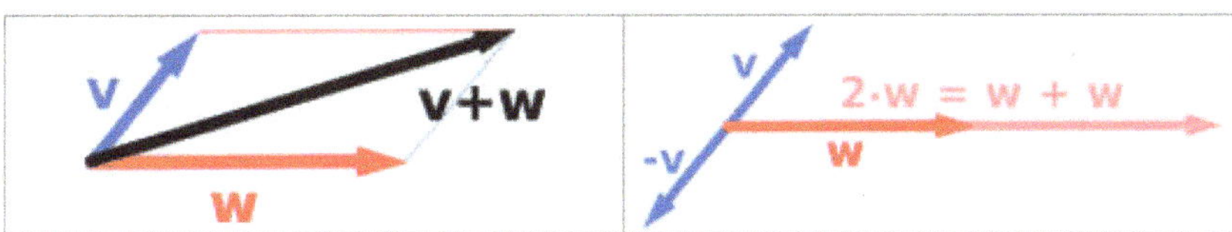

Second Example: Ordered Pairs of Numbers

A second key example of a vector space is provided by pairs of real numbers x and y. (The order of the components x and y is significant, so such a pair is also called an ordered pair.) Such a pair is written as (x, y). The sum of two such pairs and multiplication of a pair with a number is defined as follows:

$(x_1, y_1) + (x_2, y_2) = (x_1 + x_2, y_1 + y_2)$

and

$a(x, y) = (ax, ay).$

The first example above reduces to this one if the arrows are represented by the pair of Cartesian coordinates of their end points.

Definition

A vector space over a field F is a set V together with two operations that satisfy the eight axioms listed below. Elements of V are commonly called vectors. Elements of F are commonly called scalars. The first operation, called vector addition or simply addition, takes any two vectors **v** and **w**and assigns to them a third vector which is commonly written as **v** + **w**, and called the sum of these two vectors. The second operation, called scalar multiplication takes any scalar a and any vector **v** and gives another vector a**v**.

In this article, vectors are distinguished from scalars by boldface.[nb 1] In the two examples above, the field is the field of the real numbers and the set of the vectors consists of the planar arrows with fixed starting point and of pairs of real numbers, respectively.

To qualify as a vector space, the set V and the operations of addition and multiplication must adhere to a number of requirements called axioms.In the list below, let **u**, **v** and **w** be arbitrary vectors in V, and a and b scalars in F.

Axiom	Meaning
Associativity of addition	$\mathbf{u} + (\mathbf{v} + \mathbf{w}) = (\mathbf{u} + \mathbf{v}) + \mathbf{w}$
Commutativity of addition	$\mathbf{u} + \mathbf{v} = \mathbf{v} + \mathbf{u}$
Identity element of addition	There exists an element $\mathbf{0} \in V$, called the *zero vector*, such that $\mathbf{v} + \mathbf{0} = \mathbf{v}$ for all $\mathbf{v} \in V$.
Inverse elements of addition	For every $\mathbf{v} \in V$, there exists an element $-\mathbf{v} \in V$, called the *additive inverse* of $\mathbf{v}$, such that $\mathbf{v} + (-\mathbf{v}) = \mathbf{0}$.
Compatibility of scalar multiplication with field multiplication	$a(b\mathbf{v}) = (ab)\mathbf{v}$ [nb 2]
Identity element of scalar multiplication	$1\mathbf{v} = \mathbf{v}$, where 1 denotes the multiplicative identity in F.
Distributivity of scalar multiplication with respect to vector addition	$a(\mathbf{u} + \mathbf{v}) = a\mathbf{u} + a\mathbf{v}$
Distributivity of scalar multiplication with respect to field addition	$(a + b)\mathbf{v} = a\mathbf{v} + b\mathbf{v}$

These axioms generalize properties of the vectors introduced in the above examples. Indeed, the result of addition of two ordered pairs (as in the second example above) does not depend on the order of the summands:

$(x_v, y_v) + (x_w, y_w) = (x_w, y_w) + (x_v, y_v).$

Likewise, in the geometric example of vectors as arrows, **v** + **w** = **w** + **v** since the parallelogram defining the sum of the vectors is independent of the order of the vectors. All other axioms can be checked in a similar manner in both examples. Thus, by disregarding the concrete nature of the particular type of vectors, the definition incorporates these two and many more examples in one notion of vector space.

Subtraction of two vectors and division by a (non-zero) scalar can be defined as

$\mathbf{v} - \mathbf{w} = \mathbf{v} + (-\mathbf{w}),$

$\mathbf{v}/a = (1/a)\mathbf{v}.$

When the scalar field F is the real numbers **R**, the vector space is called a real vector space. When the scalar field is the complex numbers, it is called a complex vector space. These two cases are the ones used most often in engineering. The general definition of a vector space allows scalars to be elements of any fixed field F. The notion is then known as an F-vector spaces or a vector space over F. A field is, essentially, a set of numbers possessing addition, subtraction, multiplication and division operations.[nb 3] For example, rational numbers also form a field.

In contrast to the intuition stemming from vectors in the plane and higher-dimensional cases, there is, in general vector spaces, no notion of nearness, angles or distances. To deal with such matters, particular types of vector spaces are introduced; see below.

Alternative Formulations and Elementary Consequences

Vector addition and scalar multiplication are operations, satisfying the closure property: **u** + **v** and a**v** are in V for all a in F, and **u**, **v** in V. Some older sources mention these properties as separate axioms.

In the parlance of abstract algebra, the first four axioms can be subsumed by requiring the set of vectors to be an abelian group under addition. The remaining axioms give this group an F-module structure. In other words, there is a ring homomorphism f from the field F into the endomorphism ring of the group of vectors. Then scalar multiplication a**v** is defined as (f(a))(**v**).

There are a number of direct consequences of the vector space axioms. Some of them derive from elementary group theory, applied to the additive group of vectors: for example the zero vector **0** of V and the additive inverse −**v** of any vector **v** are unique. Other properties follow from the distributive law, for example a**v** equals **0** if and only if a equals 0 or **v** equals **0**.

Vector spaces stem from affine geometry via the introduction of coordinates in the plane or three-dimensional space. Around 1636, Descartes and Fermat founded analytic geometry by equating solutions to an equation of two variables with points on a plane curve. In 1804, to achieve geometric solutions without using coordinates, Bolzano introduced certain operations on points, lines and planes, which are predecessors of vectors. His work was then used in the conception of barycentric coordinates by Möbius in 1827. In 1828 C. V. Mourey suggested the existence of an algebra surpassing not only ordinary algebra but also two-dimensional algebra created by him searching a geometrical interpretation of complex numbers.

The definition of vectors was founded on Bellavitis' notion of the bipoint, an oriented segment of which one end is the origin and the other a target, then further elaborated with the presentation of complex numbers by Argand and Hamilton and the introduction of quaternions and biquaternions by the latter. They are elements in $\mathbf{R}^2$, $\mathbf{R}^4$, and $\mathbf{R}^8$; their treatment as linear combinations can be traced back to Laguerre in 1867, who also defined systems of linear equations.

In 1857, Cayley introduced matrix notation, which allows for a harmonization and simplification of linear maps. Around the same time, Grassmann studied the barycentric calculus initiated by Möbius. He envisaged sets of abstract objects endowed with operations. In his work, the concepts of linear independence and dimension, as well as scalar products, are present. In fact, Grassmann's 1844 work exceeds the framework of vector spaces, since his consideration of multiplication led him to what are today called algebras. Peano was the first to give the modern definition of vector spaces and linear maps in 1888.

An important development of vector spaces is due to the construction of function spaces by Lebesgue. This was later formalized by Banach and Hilbert, around 1920. At that time, algebra and the new field of functional analysis began to interact, notably with key concepts such as spaces of p-integrable functions and Hilbert spaces. Vector spaces, including infinite-dimensional ones, then became a firmly established notion, and many mathematical branches started making use of this concept.

Examples

Coordinate spaces

The simplest example of a vector space over a field F is the field itself, equipped with its standard addition and multiplication. More generally, a vector space can be composed of n-tuples (sequences of length n) of elements of F, such as

$(a_1, a_2, ..., a_n)$, where each a_i is an element of F.

A vector space composed of all the n-tuples of a field F is known as a coordinate space, usually denoted F^n. The case n = 1 is the above-mentioned simplest example, in which the field F is also regarded as a vector space over itself. The case F = **R** and n = 2 was discussed in the introduction above.

Complex Numbers and Other Field Extensions

The set of complex numbers **C**, i.e., numbers that can be written in the form x + iy for real numbers x and y where i is the imaginary unit, form a vector space over the reals with the usual addition and multiplication: (x + iy) + (a + ib) = (x + a) + i(y + b) and c · (x + iy) = (c · x) + i(c · y) for real numbers x, y, a, b and c. The various axioms of a vector space follow from the fact that the same rules hold for complex number arithmetic.

In fact, the example of complex numbers is essentially the same (i.e., it is isomorphic) to the vector space of ordered pairs of real numbers mentioned above: if we think of the complex number x + i y as representing the ordered pair (x, y) in the complex plane then we see that the rules for sum and scalar product correspond exactly to those in the earlier example.

More generally, field extensions provide another class of examples of vector spaces, particularly in algebra and algebraic number theory: a field Fcontaining a smaller field E is an E-vector space, by the given multiplication and addition operations of F. For example, the complex numbers are a vector space over **R**, and the field extension $\mathbf{Q}(i\sqrt{5})$ is a vector space over **Q**.

Function Spaces

Functions from any fixed set Ω to a field F also form vector spaces, by performing addition and scalar multiplication pointwise. That is, the sum of two functions f and g is the function (f + g) given by

(f + g)(w) = f(w) + g(w),

and similarly for multiplication. Such function spaces occur in many geometric situations, when Ω is the real line or an interval, or other subsets of **R**. Many notions in topology and analysis, such as continuity, integrability or differentiability are well-behaved with respect to linearity: sums and scalar multiples of functions possessing such a property still have that property. Therefore, the set of such functions are vector spaces. They are studied in greater detail using the methods of functional analysis, see below. Algebraic constraints also yield vector spaces: the vector space F[x]is given by polynomial functions: $f(x) = r_0 + r_1x + ... + r_{n-1}x^{n-1} + r_nx^n$, where the coefficients $r_0, ..., r_n$ are in F.

Linear Equations

Systems of homogeneous linear equations are closely tied to vector spaces. For example, the solutions of

a	+	3b	+	c	= 0
4a	+	2b	+	2c	= 0

are given by triples with arbitrary a, b = a/2, and c = −5a/2. They form a vector space: sums and scalar multiples of such triples still satisfy the same ratios of the three variables; thus they are solutions, too. Matrices can be used to condense multiple linear equations as above into one vector equation, namely

$A\boldsymbol{x} = \boldsymbol{0}$,

where $A = \begin{bmatrix} 1 & 3 & 1 \\ 4 & 2 & 2 \end{bmatrix}$ is the matrix containing the coefficients of the given equations, **x** is the vector (*a*, *b*, *c*), *A***x** denotes the matrix product, and **0** = (0, 0) is the zero vector. In a similar vein, the solutions of homogeneous *linear differential equations* form vector spaces. For example,

$f''(x) + 2f'(x) + f(x) = 0$

yields $f(x) = a e^{-x} + bx e^{-x}$, where a and b are arbitrary constants, and e^x is the natural exponential function.

Basis and Dimension

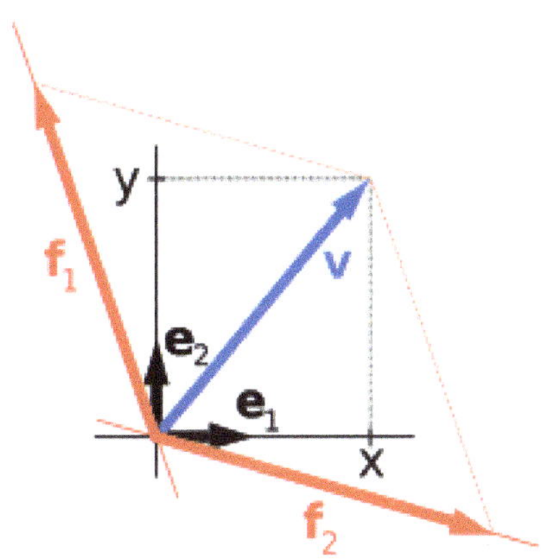

A vector **v** in $\mathbf{R}^2$ (blue) expressed in terms of different bases: using the standard basis of $\mathbf{R}^2$ $\mathbf{v} = x\mathbf{e}_1 + y\mathbf{e}_2$ (black), and using a different, non-orthogonal basis: $\mathbf{v} = \mathbf{f}_1 + \mathbf{f}_2$ (red).

Bases allow one to represent vectors by a sequence of scalars called coordinates or components. A basis is a (finite or infinite) set $B = \{\mathbf{b}_i\}_{i \in I}$ of vectors $\mathbf{b}_i$, for convenience often indexed by some index set I, that spans the whole space and is linearly independent. "Spanning the whole space" means that any vector **v** can be expressed as a finite sum (called a linear combination) of the basis elements:

$$\mathbf{v} = a_1 \mathbf{b}_{i_1} + a_2 \mathbf{b}_{i_2} + \cdots + a_n \mathbf{b}_{i_n},$$

where the a_k are scalars, called the coordinates (or the components) of the vector **v** with respect to the basis B, and $\mathbf{b}_{ik}$ ($k = 1, ..., n$) elements of B. Linear independence means that the coordinates a_k are uniquely determined for any vector in the vector space.

For example, the coordinate vectors $\mathbf{e}_1 = (1, 0, ..., 0)$, $\mathbf{e}_2 = (0, 1, 0, ..., 0)$, to $\mathbf{e}_n = (0, 0, ..., 0, 1)$, form a basis of F^n, called the standard basis, since any vector $(x_1, x_2, ..., x_n)$ can be uniquely expressed as a linear combination of these vectors:

$(x_1, x_2, ..., x_n) = x_1(1, 0, ..., 0) + x_2(0, 1, 0, ..., 0) + ... + x_n(0, ..., 0, 1) = x_1\mathbf{e}_1 + x_2\mathbf{e}_2 + ... + x_n\mathbf{e}_n$.

The corresponding coordinates $x_1, x_2, ..., x_n$ are just the Cartesian coordinates of the vector.

Every vector space has a basis. This follows from Zorn's lemma, an equivalent formulation of the Axiom of Choice. Given the other axioms of Zermelo–Fraenkel set theory, the existence of bases is equivalent to the axiom of choice. The ultrafilter lemma, which is weaker than the axiom of choice, implies that all bases of a given vector space have the same number of elements, or cardinality (cf. Dimension theorem for vector spaces). It is called the dimension of the vector space, denoted dim V. If the space is spanned by finitely many vectors, the above statements can be proven without such fundamental input from set theory.

The dimension of the coordinate space F^n is n, by the basis exhibited above. The dimension of the polynomial ring F[x] introduced above is countably infinite, a basis is given by $1, x, x^2, ...$ A fortiori, the dimension of more general function spaces, such as the space of functions on some (bounded or unbounded) interval, is infinite.[nb 4] Under suitable regularity assumptions on the coefficients involved, the dimension of the solution space of a homogeneous ordinary differential equation equals the degree of the equation. For example, the solution space for the above equation is generated by c^{-x} and xc^{-x}. These two functions are linearly independent over **R**, so the dimension of this space is two, as is the degree of the equation.

A field extension over the rationals **Q** can be thought of as a vector space over **Q** (by defining vector addition as field addition, defining scalar multiplication as field multiplication by elements of **Q**, and otherwise ignoring the field multiplication). The dimension (or degree) of the field extension **Q**(α) over **Q** depends on α. If α satisfies some polynomial equation

$$q_n \alpha^n + q_{n-1} \alpha^{n-1} + \ldots + q_0 = 0$$

with rational coefficients $q_n, ..., q_0$ (in other words, if α is algebraic), the dimension is finite. More precisely, it equals the degree of the minimal polynomial having α as a root. For example, the complex numbers **C** are a two-dimensional real vector space, generated by 1 and the imaginary unit i. The latter satisfies $i^2 + 1 = 0$, an equation of degree two. Thus, **C** is a two-dimensional **R**-vector space (and, as any field, one-dimensional as a vector space over itself, **C**). If α is not algebraic, the dimension of **Q**(α) over **Q** is infinite. For instance, for α = π there is no such equation, in other words π is transcendental.

Linear Maps and Matrices

The relation of two vector spaces can be expressed by linear map or linear transformation. They are functions that reflect the vector space structure—i.e., they preserve sums and scalar multiplication:

$f(\mathbf{x} + \mathbf{y}) = f(\mathbf{x}) + f(\mathbf{y})$ and $f(a \cdot \mathbf{x}) = a \cdot f(\mathbf{x})$ for all **x** and **y** in V, all a in F.

An isomorphism is a linear map $f : V \to W$ such that there exists an inverse map $g : W \to V$, which is a map such that the two possible compositions $f \circ g : W \to W$ and $g \circ f : V \to V$ are identity maps. Equivalently, f is both one-to-one (injective) and onto (surjective). If there exists an isomorphism between V and W, the two spaces are said to be isomorphic; they are then essentially identical as vector spaces, since all identities holding in V are, via f, transported to similar ones in W, and vice versa via g.

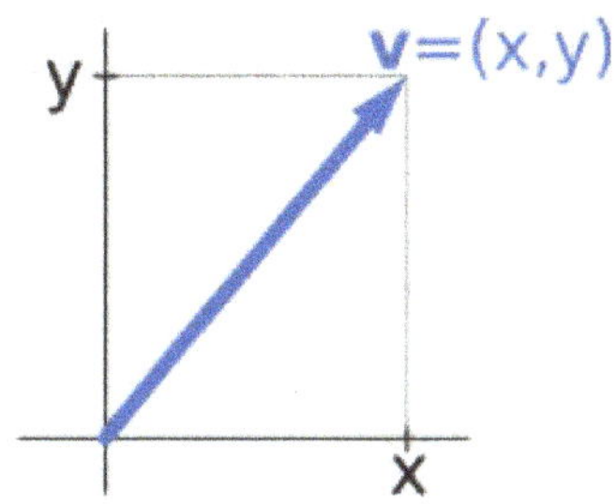

Describing an arrow vector **v** by its coordinates x and y yields an isomorphism of vector spaces.

For example, the "arrows in the plane" and "ordered pairs of numbers" vector spaces in the introduction are isomorphic: a planar arrow **v** departing at the origin of some (fixed) coordinate system can be expressed as an ordered pair by considering the x- and y-component of the arrow, as shown in the image at the right. Conversely, given a pair (x, y), the arrow going by x to the right (or to the left, if x is negative), and y up (down, if y is negative) turns back the arrow **v**.

Linear maps V → W between two vector spaces form a vector space $\mathrm{Hom}_F(V, W)$, also denoted L(V, W). The space of linear maps from V to F is called the dual vector space, denoted V^*. Via the injective natural map $V \to V^{**}$, any vector space can be embedded into its bidual; the map is an isomorphism if and only if the space is finite-dimensional.

Once a basis of V is chosen, linear maps $f : V \to W$ are completely determined by specifying the images of the basis vectors, because any element of V is expressed uniquely as a linear combination of them. If dim V = dim W, a 1-to-1 correspondence between fixed bases of V and W gives rise to a linear map that maps any basis element of V to the corresponding basis element of W. It is an isomorphism, by its very definition. Therefore, two vector spaces are isomorphic if their dimensions agree and vice versa. Another way to express this is that any vector space is completely classified (up to isomorphism) by its dimension, a single number. In particular, any n-dimensional F-vector space V is isomorphic to F^n. There is, however, no "canonical" or preferred isomorphism; actually an isomorphism $\varphi : F^n \to V$ is equivalent to the choice of a basis of V, by mapping the standard basis of F^n to V, via φ. The freedom of choosing a convenient basis is particularly useful in the infinite-dimensional context, see below.

Matrices

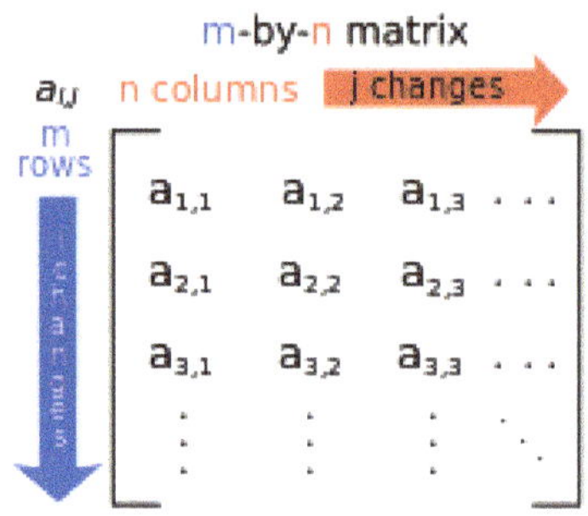

A typical matrix

Matrices are a useful notion to encode linear maps. They are written as a rectangular array of scalars as in the image at the right. Any m-by-n matrix A gives rise to a linear map from F^n to F^m, by the following

$$\mathbf{x} = (x_1, x_2, \cdots, x_n) \mapsto \left(\sum_{j=1}^{n} a_{1j}x_j, \sum_{j=1}^{n} a_{2j}x_j, \cdots, \sum_{j=1}^{n} a_{mj}x_j\right)$$, where $\sum$ donotes summation,

or, using the matrix multiplication of the matrix A with the coordinate vector **x**:

$x \mapsto Ax$.

Moreover, after choosing bases of V and W, any linear map $f : V \to W$ is uniquely represented by a matrix via this assignment.

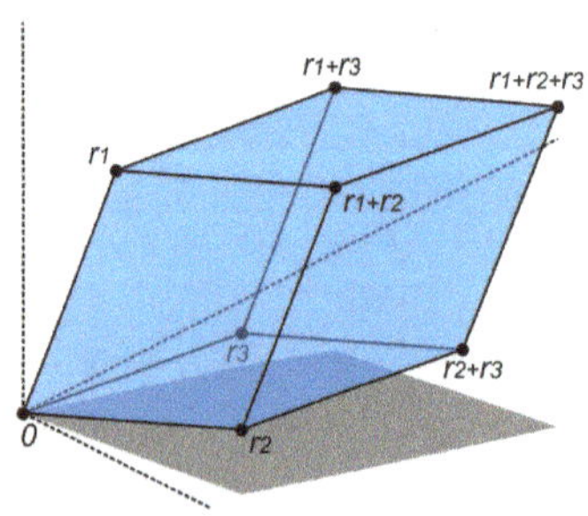

The volume of this parallelepiped is the absolute value of the determinant of the 3-by-3 matrix formed by the vectors r_1, r_2, and r_3.

The determinant det (A) of a square matrix A is a scalar that tells whether the associated map is an isomorphism or not: to be so it is sufficient and necessary that the determinant is nonzero. The linear transformation of $\mathbf{R}^n$ corresponding to a real n-by-n matrix is orientation preserving if and only if its determinant is positive.

Eigenvalues and Eigenvectors

Endomorphisms, linear maps $f : V \to V$, are particularly important since in this case vectors $\mathbf{v}$ can be compared with their image under f, $f(\mathbf{v})$. Any nonzero vector $\mathbf{v}$ satisfying $\lambda\mathbf{v} = f(\mathbf{v})$, where λ is a scalar, is called an eigenvector of f with eigenvalue λ.[nb 5] Equivalently, $\mathbf{v}$ is an element of the kernel of the difference $f - \lambda \cdot \mathrm{Id}$ (where Id is the identity map $V \to V$). If V is finite-dimensional, this can be rephrased using determinants: f having eigenvalue λ is equivalent to

$$\det(f - \lambda \cdot \mathrm{Id}) = 0.$$

By spelling out the definition of the determinant, the expression on the left hand side can be seen to be a polynomial function in λ, called the characteristic polynomial of f. If the field F is large enough to contain a zero of this polynomial (which automatically happens for F algebraically closed, such as $F = \mathbf{C}$) any linear map has at least one eigenvector. The vector space V may or may not possess an eigenbasis, a basis consisting of eigenvectors. This phenomenon is governed by the Jordan canonical form of the map.[nb 6] The set of all eigenvectors corresponding to a particular eigenvalue of f forms a vector space known as the eigenspace corresponding to the eigenvalue (and f) in question. To achieve the spectral theorem, the corresponding statement in the infinite-dimensional case, the machinery of functional analysis is needed, see below.

BASIC CONSTRUCTIONS

In addition to the above concrete examples, there are a number of standard linear algebraic constructions that yield vector spaces related to given ones. In addition to the definitions given below, they are also characterized by universal properties, which determine an object X by specifying the linear maps from X to any other vector space.

Subspaces and Quotient Spaces

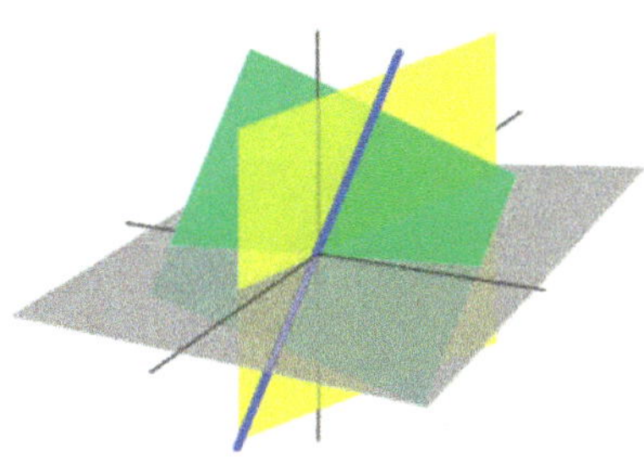

A line passing through the origin (blue, thick) in $\mathbf{R}^3$ is a linear subspace. It is the intersection of two planes (green and yellow).

A nonempty subset W of a vector space V that is closed under addition and scalar multiplication (and therefore contains the **0**-vector of V) is called a linear subspace of V, or simply a subspace of V, when the ambient space is unambiguously a vector space. [nb 7] Subspaces of V are vector spaces (over the same field) in their own right. The intersection of all subspaces containing a given set S of vectors is called its span, and it is the smallest subspace of V containing the set S. Expressed in terms of elements, the span is the subspace consisting of all the linear combinations of elements of S.

A linear subspace of dimension 1 is a **vector line**. A linear subspace of dimension 2 is a **vector plane**. A linear subspace that contains all elements but one of a basis of the ambient space is a **vector hyperplane**. In a vector space of finite dimension n, a vector hyperplane is thus a subspace of dimension n – 1.

The counterpart to subspaces are quotient vector spaces. Given any subspace $W \subset V$, the quotient space V/W ("V modulo W") is defined as follows: as a set, it consists of $\mathbf{v} + W = \{\mathbf{v} + \mathbf{w} : \mathbf{w} \in W\}$, where **v** is an arbitrary vector in V. The sum of two such elements $\mathbf{v}_1 + W$ and $\mathbf{v}_2 + W$ is $(\mathbf{v}_1 + \mathbf{v}_2) + W$, and scalar multiplication is given by $a \cdot (\mathbf{v} + W) = (a \cdot \mathbf{v}) + W$. The key point in this definition is that $\mathbf{v}_1 + W = \mathbf{v}_2 + W$ if and only if the difference of $\mathbf{v}_1$ and $\mathbf{v}_2$ lies in W.[nb 8] This way, the quotient space "forgets" information that is contained in the subspace W.

The kernel ker(f) of a linear map $f : V \to W$ consists of vectors **v** that are mapped to **0** in W. Both kernel and image $\mathrm{im}(f) = \{f(\mathbf{v}) : \mathbf{v} \in V\}$ are subspaces of V and W, respectively. The existence of kernels and images is part of the statement that the category of vector spaces (over a fixed field F) is an abelian category, i.e. a corpus of mathematical objects and structure-preserving maps between them (a category) that behaves much like the category of abelian groups. Because of this, many statements such as the first isomorphism theorem (also called rank–nullity theorem in matrix-related terms)

$V / \ker(f) \equiv \mathrm{im}(f).$

and the second and third isomorphism theorem can be formulated and proven in a way very similar to the corresponding statements for groups.

An important example is the kernel of a linear map $\mathbf{x} \mapsto A\mathbf{x}$ for some fixed matrix A, as above. The kernel of this map is the subspace of vectors **x** such that $A\mathbf{x} = 0$, which is precisely the set of solutions to the system of homogeneous linear equations belonging to A. This concept also extends to linear differential equations

$$a_0 f + a_1 \frac{df}{dx} + a_2 \frac{d^2 f}{dx^2} + \cdots + a_n \frac{d^n f}{dx^n} = 0,$$ where the coefficients a_i are functions in x, too.

In the corresponding map

$$f \mapsto D(f) = \sum_{i=0}^{n} a_i \frac{d^i f}{dx^i},$$

the derivatives of the function f appear linearly (as opposed to $f''(x)^2$, for example). Since differentiation is a linear procedure (i.e., $(f + g)' = f' + g'$ and $(c \cdot f)' = c \cdot f'$ for a constant c) this assignment is linear, called a linear differential operator. In particular, the solutions to the differential equation $D(f) = 0$ form a vector space (over **R** or **C**).

Direct Product and Direct Sum

The direct product of vector spaces and the direct sum of vector spaces are two ways of combining an indexed family of vector spaces into a new vector space.

The direct product $\prod_{i\in I} V_i$ of a family of vector spaces V_i consists of the set of all tuples $(\mathbf{v}_i)_{i \in I}$, which specify for each index i in some index set I an element $\mathbf{v}_i$ of V_i. Addition and scalar multiplication is performed componentwise. A variant of this construction is the direct sum $\bigoplus_{i\in I} V_i$ (also called coproduct and denoted $\coprod_{i\in I} V_i$), where only tuples with finitely many nonzero vectors are allowed. If the index set I is finite, the two constructions agree, but in general they are different.

Tensor Product

The tensor product $V \otimes_F W$, or simply $V \otimes W$, of two vector spaces V and W is one of the central notions of multilinear algebra which deals with extending notions such as linear maps to several variables. A map $g : V \times W \to X$ is called bilinear if g is linear in both variables **v** and **w**. That is to say, for fixed **w** the map $\mathbf{v} \mapsto g(\mathbf{v}, \mathbf{w})$ is linear in the sense above and likewise for fixed **v**.

The tensor product is a particular vector space that is a universal recipient of bilinear maps g, as follows. It is defined as the vector space consisting of finite (formal) sums of symbols called tensors

$\mathbf{v}_1 \otimes \mathbf{w}_1 + \mathbf{v}_2 \otimes \mathbf{w}_2 + \ldots + \mathbf{v}_n \otimes \mathbf{w}_n$,

subject to the rules

$a \cdot (v \otimes w) = (a \cdot v) \otimes w = v \otimes (a \cdot w)$, where a is a scalar,

$(v_1 + v_2) \otimes w = v_1 \otimes w + v_2 \otimes w$, and

$v \otimes (w_1 + w_2) = v \otimes w_1 + v \otimes w_2$.

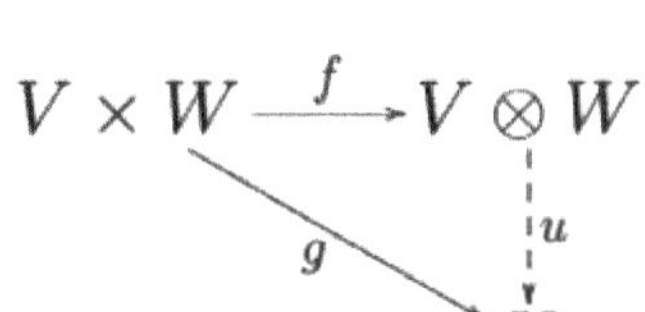

Commutative diagram depicting the universal property of the tensor product.

These rules ensure that the map f from the $V \times W$ to $V \otimes W$ that maps a tuple (**v**, **w**) to $\mathbf{v} \otimes \mathbf{w}$ is bilinear. The universality states that given any vector space X and any bilinear map $g : V \times W \to X$, there exists a unique map u, shown in the diagram with a dotted arrow, whose composition with f equals g: $u(\mathbf{v} \otimes \mathbf{w}) = g(\mathbf{v}, \mathbf{w})$. This is called the universal property of the tensor product, an instance of the method—much used in advanced abstract algebra—to indirectly define objects by specifying maps from or to this object.

Vector Spaces With Additional Structure

From the point of view of linear algebra, vector spaces are completely understood insofar as any vector space is characterized, up to isomorphism, by its dimension. However, vector spaces per se do not offer a framework to deal with the question—crucial to analysis—whether a sequence of functions converges to another function. Likewise, linear algebra is not adapted to deal with infinite series, since the addition operation allows only finitely many terms to be added. Therefore, the needs of functional analysis require considering additional structures.

A vector space may be given a partial order $\leq$, under which some vectors can be compared. For example, n-dimensional real space $\mathbf{R}^n$ can be ordered by comparing its vectors componentwise. Ordered vector spaces, for example Riesz spaces, are fundamental to Lebesgue integration, which relies on the ability to express a function as a difference of two positive functions

$f = f^+ - f^-$,

where f^+ denotes the positive part of f and f^- the negative part.

Topological Vector Spaces

Convergence questions are treated by considering vector spaces V carrying a compatible topology, a structure that allows one to talk about elements being close to each other. Compatible here means that addition and scalar multiplication have to be continuous maps. Roughly, if x and y in V, and a in F vary by a bounded amount, then so do x + y and ax.[nb 9] To make sense of specifying the amount a scalar changes, the field F also has to carry a topology in this context; a common choice are the reals or the complex numbers.

In such topological vector spaces one can consider series of vectors. The infinite sum

$$\sum_{i=0}^{\infty} f_i$$

denotes the limit of the corresponding finite partial sums of the sequence $(f_i)_{i\in N}$ of elements of V. For example, the f_i could be (real or complex) functions belonging to some function space V, in which case the series is a function series. The mode of convergence of the series depends on the topology imposed on the function space. In such cases, pointwise convergence and uniform convergence are two prominent examples.

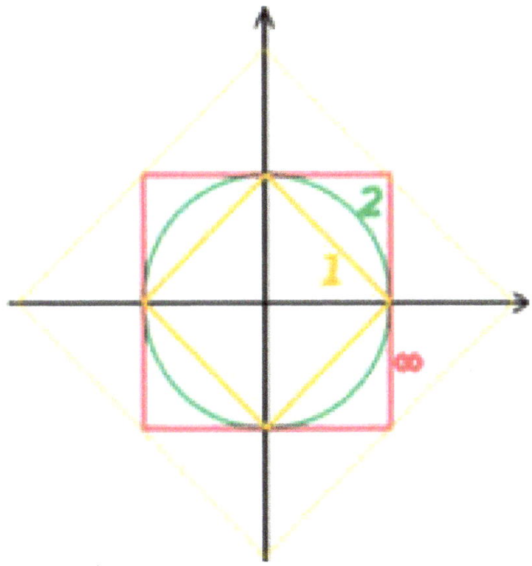

Unit "spheres" in $\mathbf{R}^2$ consist of plane vectors of norm 1. Depicted are the unit spheres in different p-norms, for p = 1, 2, and ∞. The bigger diamond depicts points of 1-norm equal to {\displaystyle {\sqrt {2}}}.

A way to ensure the existence of limits of certain infinite series is to restrict attention to spaces where any Cauchy sequence has a limit; such a vector space is called complete. Roughly, a vector space is complete provided that it contains all necessary limits. For example, the vector space of polynomials on the unit interval [0,1], equipped with the topology of uniform convergence is not complete because any continuous function on [0,1] can be uniformly approximated by a sequence of polynomials, by the Weierstrass approximation theorem. In contrast, the space of all continuous functions on [0,1] with the same topology is complete. A norm gives rise to a topology by defining that a sequence of vectors $\mathbf{v}_n$ converges to $\mathbf{v}$ if and only if

$$\lim_{n\to\infty} |\mathbf{v}_n - \mathbf{v}| = 0.$$

Banach and Hilbert spaces are complete topological vector spaces whose topologies are given, respectively, by a norm and an inner product. Their study—a key piece of functional analysis—focusses on infinite-dimensional vector spaces, since all norms on finite-dimensional topological vector spaces give rise to the same notion of convergence. The image at the right shows the equivalence of the 1-norm and ∞-norm on $\mathbf{R}^2$: as the unit "balls" enclose each other, a sequence converges to zero in one norm if and only if it so does in the other norm. In the infinite-dimensional case, however, there will generally be inequivalent topologies, which makes the study of topological vector spaces richer than that of vector spaces without additional data.

From a conceptual point of view, all notions related to topological vector spaces should match the topology. For example, instead of considering all linear maps (also called functionals) V → W, maps between topological vector spaces are required to be continuous. In particular, the (topological) dual space V* consists of continuous functionals

$V \rightarrow \mathbf{R}$ (or to $\mathbf{C}$). The fundamental Hahn–Banach theorem is concerned with separating subspaces of appropriate topological vector spaces by continuous functionals.

Banach Spaces

Banach spaces, introduced by Stefan Banach, are complete normed vector spaces. A first example is the vector space ℓ^p consisting of infinite vectors with real entries $\mathbf{x} = (x_1, x_2, ...)$ whose p-norm ($1 \leq p \leq \infty$) given by

$$|\mathbf{x}|_p := \left(\sum_i |x_i|^p\right)^{1/p} \text{ for } p < \infty \text{ and } |\mathbf{x}|_\infty := \sup_i |x_i|$$

is finite. The topologies on the infinite-dimensional space ℓp are inequivalent for different p. E.g. the sequence of vectors $X_n = (2^{-n}, 2^{-n}, ..., 2^{-n}, 0, 0, ...)$, i.e. the first 2^n components are 2^{-n}, the following ones are 0, converges to the zero vector for $p = \infty$, but does not for $p = 1$:

$$|x_n|_\infty = \sup(2^{-n}, 0) = 2^{-n} \rightarrow 0, \text{ but } |x_n|_1 = \sum_{i=1}^{2^n} 2^{-n} = 2^n \cdot 2^{-n} = 1.$$

More generally than sequences of real numbers, functions $f\colon \Omega \rightarrow R$ are endowed with a norm that replaces the above sum by the Lebesgue integral

$$|f|_p := \left(\int_\Omega |f(x)|^p \, dx\right)^{1/p}.$$

The space of integrable functions on a given domain Ω (for example an interval) satisfying $|f|_p < \infty$, and equipped with this norm are called Lebesgue spaces, denoted $L^p(\Omega)$.[nb 10] These spaces are complete. (If one uses the Riemann integral instead, the space is not complete, which may be seen as a justification for Lebesgue's integration theory.[nb 11]) Concretely this means that for any sequence of Lebesgue-integrable functions $f_1, f_2, ...$ with $|f_n|_p < \infty$, satisfying the condition

$$\lim_{k,\, n\rightarrow\infty} \int_\Omega |f_k(x) - f_n(x)|^p \, dx = 0$$

there exists a function f(x) belonging to the vector space $L^p(\Omega)$ such that

$$\lim_{k\rightarrow\infty} \int_\Omega |f(x) - f_k(x)|^p \, dx = 0.$$

Imposing boundedness conditions not only on the function, but also on its derivatives leads to Sobolev spaces.

Hilbert Spaces

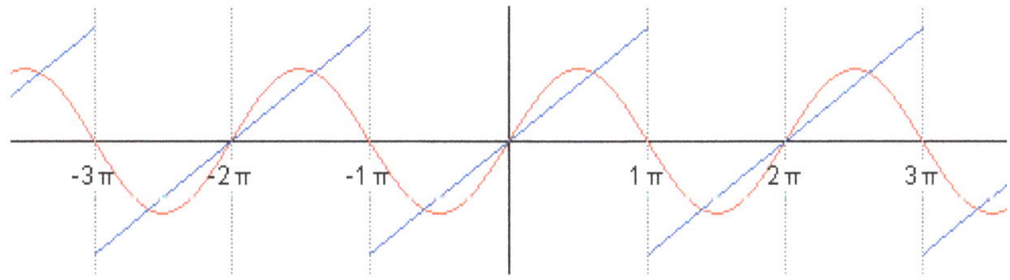

The succeeding snapshots show summation of 1 to 5 terms in approximating a periodic function (blue) by finite sum of sine functions (red).

Complete inner product spaces are known as Hilbert spaces, in honor of David Hilbert. The Hilbert space $L^2(\Omega)$, with inner product given by

$$\langle f\, ,\, g\rangle = \int_\Omega f(x)\overline{g(x)}\, dx,$$

where $\overline{g(x)}$ denotes the complex conjugate of g(x), is a key case.

By definition, in a Hilbert space any Cauchy sequence converges to a limit. Conversely, finding a sequence of functions f_n with desirable properties that approximates a given limit function, is equally crucial. Early analysis, in the guise of the Taylor approximation, established an approximation of differentiable functions f by polynomials. By the Stone–Weierstrass theorem, every continuous function on [a, b] can be approximated as closely as desired by a polynomial. A similar approximation technique by trigonometric functions is commonly called Fourier expansion, and is much applied in engineering, see below. More generally, and more conceptually, the theorem yields a simple description of what "basic functions", or, in abstract Hilbert spaces, what basic vectors suffice to generate a Hilbert space H, in the sense that the closure of their span (i.e., finite linear combinations and limits of those) is the whole space. Such a set of functions is called a basis of H, its cardinality is known as the Hilbert space dimension.[nb 13] Not only does the theorem exhibit suitable basis functions as sufficient for approximation purposes, but together with the Gram-Schmidt process, it enables one to construct a basis of orthogonal vectors. Such orthogonal bases are the Hilbert space generalization of the coordinate axes in finite-dimensional Euclidean space.

The solutions to various differential equations can be interpreted in terms of Hilbert spaces. For example, a great many fields in physics and engineering lead to such equations and frequently solutions with particular physical properties are used as basis functions, often orthogonal. As an example from physics, the time-dependent Schrödinger equation in quantum mechanics describes the change of physical properties in time by means of a partial differential equation, whose solutions are called wavefunctions. Definite values for physical properties such as energy, or momentum, correspond to eigenvalues of a certain (linear) differential operator and the associated wavefunctions are called eigenstates. The spectral theorem decomposes a linear compact operator acting on functions in terms of these eigenfunctions and their eigenvalues.

Algebras over Fields

General vector spaces do not possess a multiplication between vectors. A vector space equipped with an additional bilinear operator defining the multiplication of two vectors is an algebra over a field.Many algebras stem from functions on some geometrical object: since functions with values in a given field can be multiplied pointwise, these entities form algebras. The Stone–Weierstrass theorem mentioned above, for example, relies on Banach algebras which are both Banach spaces and algebras.

Commutative algebra makes great use of rings of polynomials in one or several variables, introduced above. Their multiplication is both commutative and associative. These rings and their quotients form the basis of algebraic geometry, because they are rings of functions of algebraic geometric objects.

Another crucial example are Lie algebras, which are neither commutative nor associative, but the failure to be so is limited by the constraints ([x, y] denotes the product of x and y):

- $[x, y] = -[y, x]$ (anticommutativity), and
- $[x, [y, z]] + [y, [z, x]] + [z, [x, y]] = 0$ (Jacobi identity).

Examples include the vector space of n-by-n matrices, with $[x, y] = xy - yx$, the commutator of two matrices, and $\mathbf{R}^3$, endowed with the cross product.

The tensor algebra T(V) is a formal way of adding products to any vector space V to obtain an algebra. As a vector space, it is spanned by symbols, called simple tensors $\mathbf{v}_1 \otimes \mathbf{v}_2 \otimes \ldots \otimes \mathbf{v}_n$, where the degree n varies.

The multiplication is given by concatenating such symbols, imposing the distributive law under addition, and requiring that scalar multiplication commute with the tensor product $\otimes$, much the same way as with the tensor product of two vector spaces introduced above. In general, there are no relations between $\mathbf{v}_1 \otimes \mathbf{v}_2$ and $\mathbf{v}_2 \otimes \mathbf{v}_1$. Forcing two such elements to be equal leads to the symmetric algebra, whereas forcing $\mathbf{v}_1 \otimes \mathbf{v}_2 = -\mathbf{v}_2 \otimes \mathbf{v}_1$ yields the exterior algebra.

When a field, F is explicitly stated, a common term used is F-algebra.

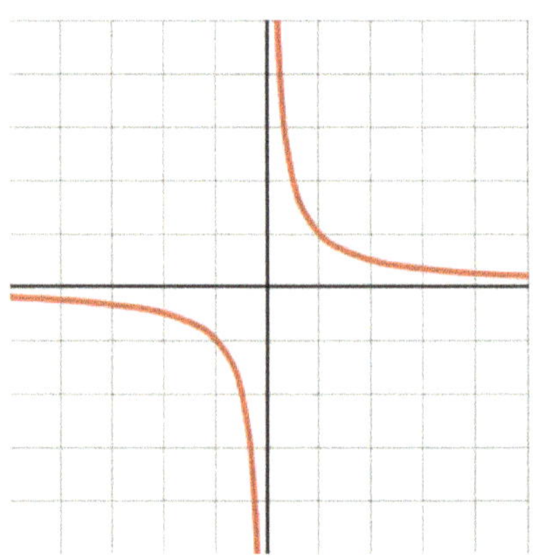

A hyperbola, given by the equation $x \cdot y = 1$. The coordinate ring of functions on this hyperbola is given by $\mathbf{R}[x, y] / (x \cdot y - 1)$, an infinite-dimensional vector space over $\mathbf{R}$.

APPLICATIONS

Vector spaces have many applications as they occur frequently in common circumstances, namely wherever functions with values in some field are involved. They provide a framework to deal with analytical and geometrical problems, or are used in the Fourier transform. This list is not exhaustive: many more applications exist, for example in optimization. The minimax theorem of game theory stating the existence of a unique payoff when all players play optimally can be formulated and proven using vector spaces methods. Representation theory fruitfully transfers the good understanding of linear algebra and vector spaces to other mathematical domains such as group theory.

Distributions

A distribution (or generalized function) is a linear map assigning a number to each "test" function, typically a smooth function with compact support, in a continuous way: in the above terminology the space of distributions is the (continuous) dual of the test function space. The latter space is endowed with a topology that takes into account not only f itself, but also all its higher derivatives. A standard example is the result of integrating a test function f over some domain Ω:

$$I(f) = \int_{\Omega} f(x)\,dx.$$

When Ω = {p}, the set consisting of a single point, this reduces to the Dirac distribution, denoted by δ, which associates to a test function f its value at the p: δ(f) = f(p). Distributions are a powerful instrument to solve differential equations. Since all standard analytic notions such as derivatives are linear, they extend naturally to the space of distributions. Therefore, the equation in question can be transferred to a distribution space, which is bigger than the underlying function space, so that more flexible methods are available for solving the equation. For example, Green's functions and fundamental solutions are usually distributions rather than proper functions, and can then be used to find solutions of the equation with prescribed boundary conditions. The found solution can then in some cases be proven to be actually a true function, and a solution to the original equation (e.g., using the Lax–Milgram theorem, a consequence of the Riesz representation theorem).

Fourier Analysis

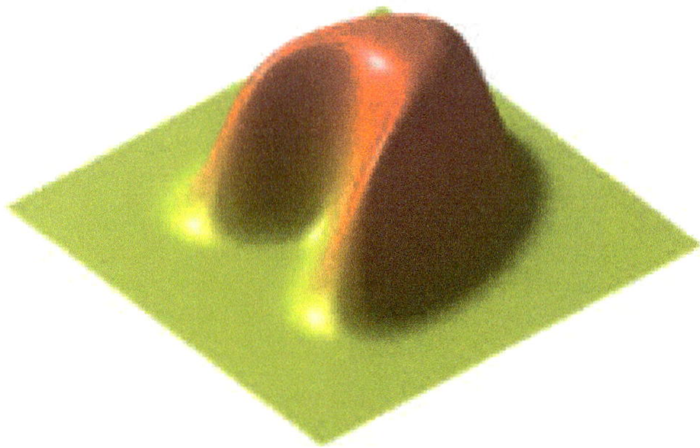

The heat equation describes the dissipation of physical properties over time, such as the decline of the temperature of a hot body placed in a colder environment (yellow depicts colder regions than red).

Resolving a periodic function into a sum of trigonometric functions forms a Fourier series, a technique much used in physics and engineering.[nb 14] The underlying vector space is usually the Hilbert space $L^2(0, 2\pi)$, for which the functions sin mx and cos mx (m an integer) form an orthogonal basis. The Fourier expansion of an L^2 function f is

$$\frac{a_0}{2} + \sum_{m=1}^{\infty} \left[a_m \cos(mx) + b_m \sin(mx) \right].$$

The coefficients a_m and b_m are called Fourier coefficients of f, and are calculated by the formulas

$$a_m = \frac{1}{\pi} \int_0^{2\pi} f(t) \cos(mt)\, dt,\ b_m = \frac{1}{\pi} \int_0^{2\pi} f(t) \sin(mt)\, dt.$$

In physical terms the function is represented as a superposition of sine waves and the coefficients give information about the function's frequency spectrum. A complex-number form of Fourier series is also commonly used. The concrete formulae above are consequences of a more general mathematical duality called Pontryagin duality. Applied to the group **R**, it yields the classical Fourier transform; an application in physics are reciprocal lattices, where the underlying group is a finite-dimensional real vector space endowed with the additional datum of a lattice encoding positions of atoms in crystals.

Fourier series are used to solve boundary value problems in partial differential equations. In 1822, Fourier first used this technique to solve the heat equation. A discrete version of the Fourier series can be used in sampling applications where the function value is known only at a finite number of equally spaced points. In this case the Fourier series is finite and its value is equal to the sampled values at all points. The set of coefficients is known as the discrete Fourier transform (DFT) of the given sample sequence. The DFT is one of the key tools of digital signal processing, a field whose applications include radar, speech encoding, image compression. The JPEG image format is an application of the closely related discrete cosine transform.

The fast Fourier transform is an algorithm for rapidly computing the discrete Fourier transform. It is used not only for calculating the Fourier coefficients but, using the convolution theorem, also for computing the convolution of two finite sequences. They in turn are applied in digital filters and as a rapid multiplication algorithm for polynomials and large integers (Schönhage-Strassen algorithm).

Differential Geometry

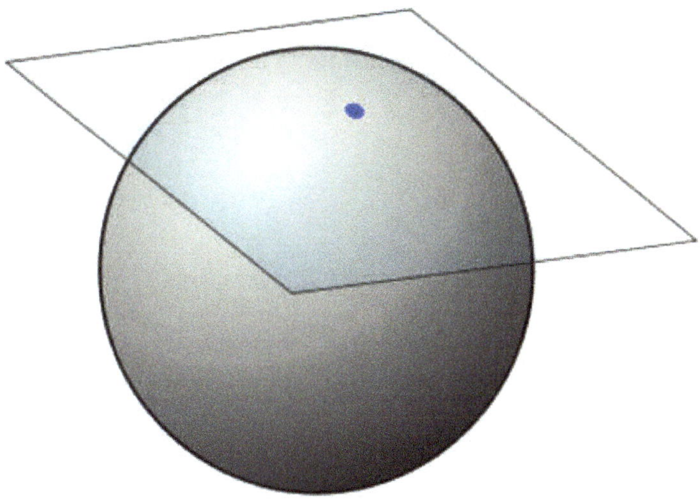

The tangent space to the 2-sphere at some point is the infinite plane touching the sphere in this point.

The tangent plane to a surface at a point is naturally a vector space whose origin is identified with the point of contact. The tangent plane is the best linear approximation, or linearization, of a surface at a point.[nb 15] Even in a three-dimensional Euclidean space, there is typically no natural way to prescribe a basis of the tangent plane, and so it is conceived of as an abstract vector space rather than a real coordinate space. The tangent space is the generalization to higher-dimensional differentiable manifolds.

Riemannian manifolds are manifolds whose tangent spaces are endowed with a suitable inner product.Derived therefrom, the Riemann curvature tensor encodes all curvatures of a manifold in one object, which finds applications in general relativity, for example, where the Einstein curvature tensor describes the matter and energy content of space-time. The tangent space of a Lie group can be given naturally the structure of a Lie algebra and can be used to classify compact Lie groups.

GENERALIZATIONS

Vector Bundles

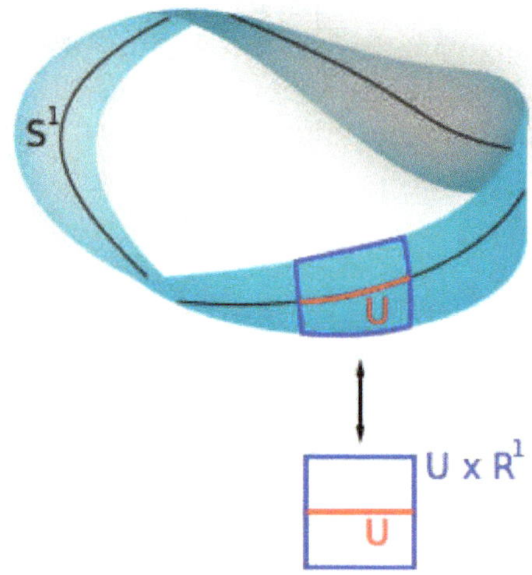

A Möbius strip. Locally, it looks likeU × **R**.

A vector bundle is a family of vector spaces parametrized continuously by a topological space X.More precisely, a vector bundle over X is a topological space E equipped with a continuous map

π: E → X

such that for every x in X, the fiber $\pi^{-1}(x)$ is a vector space. The case dim V = 1 is called a line bundle. For any vector space V, the projection X × V → X makes the product X × V into a "trivial" vector bundle. Vector bundles over X are required to be locally a product of X and some (fixed) vector space V: for every x in X, there is a neighborhood U of x such that the restriction of π to $\pi^{-1}(U)$ is isomorphic[nb 16] to the trivial bundle U × V → U. Despite their locally trivial character, vector bundles may (depending on the shape of the underlying space X) be "twisted" in the large (i.e., the bundle need not be (globally isomorphic to) the trivial bundle X × V). For example, the Möbius strip can be seen as a line bundle over the circle S^1 (by identifying open intervals with the real line). It is, however, different from the cylinder S^1 × **R**, because the latter is orientable whereas the former is not.

Properties of certain vector bundles provide information about the underlying topological space. For example, the tangent bundle consists of the collection of tangent spaces parametrized by the points of a differentiable manifold. The tangent bundle of the circle S^1 is globally isomorphic to S^1 × **R**, since there is a global nonzero vector field on S^1. In contrast, by the hairy ball theorem, there is no (tangent) vector field on the 2-sphere S^2 which is everywhere nonzero. K-theory studies the isomorphism classes of all vector bundles over some topological space. In addition to deepening topological and geometrical insight, it has purely algebraic consequences, such as the classification of finite-dimensional real division algebras: **R**, **C**, the quaternions **H** and the octonions **O**.

The cotangent bundle of a differentiable manifold consists, at every point of the manifold, of the dual of the tangent space, the cotangent space. Sections of that bundle are known as differential one-forms.

Modules

Modules are to rings what vector spaces are to fields: the same axioms, applied to a ring R instead of a field F, yield modules. The theory of modules, compared to that of vector spaces, is complicated by the presence of ring elements that do not have multiplicative inverses. For example, modules need not have bases, as the **Z**-module (i.e., abelian group) **Z**/2**Z** shows; those modules that do (including all vector spaces) are known as free modules. Nevertheless, a vector space can be compactly defined as a module over a ring which is a field with the elements being called vectors. Some authors use the term vector space to mean modules over a division ring. The algebro-geometric interpretation of commutative rings via their spectrum allows the development of concepts such as locally free modules, the algebraic counterpart to vector bundles.

Affine and Projective Spaces

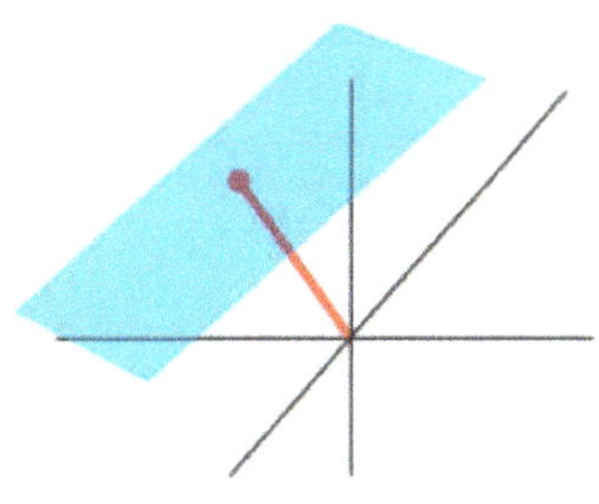

An affine plane (light blue) in $\mathbf{R}^3$. It is a two-dimensional subspace shifted by a vector **x** (red).

Roughly, affine spaces are vector spaces whose origins are not specified. More precisely, an affine space is a set with a free transitive vector space action. In particular, a vector space is an affine space over itself, by the map

$V \times V \rightarrow V, (\mathbf{v}, \mathbf{a}) \mapsto \mathbf{a} + \mathbf{v}$.

If W is a vector space, then an affine subspace is a subset of W obtained by translating a linear subspace V by a fixed vector $\mathbf{x} \in W$; this space is denoted by $\mathbf{x} + V$ (it is a coset of V in W) and consists of all vectors of the form $\mathbf{x} + \mathbf{v}$ for $\mathbf{v} \in V$. An important example is the space of solutions of a system of inhomogeneous linear equations

$A\mathbf{x} = \mathbf{b}$

generalizing the homogeneous case $\mathbf{b} = 0$ above. The space of solutions is the affine subspace $\mathbf{x} +$ Vwhere **x** is a particular solution of the equation, and V is the space of solutions of the homogeneous equation (the nullspace of A).

The set of one-dimensional subspaces of a fixed finite-dimensional vector space V is known as projective space; it may be used to formalize the idea of parallel lines intersecting at infinity. Grassmannians and flag manifolds generalize this by parametrizing linear subspaces of fixed dimension k and flags of subspaces, respectively.

REFERENCES

1. Artin, Michael (1991), Algebra, Prentice Hall, ISBN 978-0-89871-510-1
2. Blass, Andreas (1984), "Existence of bases implies the axiom of choice", Axiomatic set theory (Boulder, Colorado, 1983), Contemporary Mathematics, 31, Providence, R.I.: American Mathematical Society, pp. 31–33, MR 763890
3. Brown, William A. (1991), Matrices and vector spaces, New York: M. Dekker, ISBN 978-0-8247-8419-5
4. Lang, Serge (1987), Linear algebra, Berlin, New York: Springer-Verlag, ISBN 978-0-387-96412-6
5. Lang, Serge (2002), Algebra, Graduate Texts in Mathematics, 211 (Revised third ed.), New York: Springer-Verlag, ISBN 978-0-387-95385-4, MR 1878556
6. Mac Lane, Saunders (1999), Algebra (3rd ed.), pp. 193–222, ISBN 0-8218-1646-2
7. Meyer, Carl D. (2000), Matrix Analysis and Applied Linear Algebra, SIAM, ISBN 978-0-89871-454-8
8. Roman, Steven (2005), Advanced Linear Algebra, Graduate Texts in Mathematics, 135 (2nd ed.), Berlin, New York: Springer-Verlag, ISBN 978-0-387-24766-3
9. Spindler, Karlheinz (1993), Abstract Algebra with Applications: Volume 1: Vector spaces and groups, CRC, ISBN 978-0-8247-9144-5
10. van der Waerden, Bartel Leendert (1993), Algebra (in German) (9th ed.), Berlin, New York: Springer-Verlag, ISBN 978-3-540-56799-8

CHAPTER 7

Geometric Algebra

Chapter Contents

Geometric algebra was initiated by W.K. Clifford over 130 years ago. It unifies all branches of physics, and has found rich applications in robotics, signal processing, ray tracing, virtual reality, computer vision, vector field processing, tracking, geographic information systems and neural computing. This tutorial explains the basics of geometric algebra, with concrete examples of the plane, of 3D space, of spacetime, and the popular conformal model.

INTRODUCTION

A geometric algebra (GA) is a Clifford algebra of a vector space over the field of real numbers endowed with a quadratic form. The term is also sometimes used as a collective term for the approach to classical, computational and relativistic geometry that applies these algebras. The product of the Clifford algebra is called the geometric product in this context, and makes it a unital associative algebra. Taking the geometric product among vectors can yield bivectors, trivectors, or general n-vectors. The addition operation combines these into general multivectors, which are the elements of the ring. This includes, among other possibilities, a well-defined formal sum of a scalar and a vector.

Geometric algebra is distinguished from Clifford algebra in general by its restriction to real numbers and its emphasis on its geometric interpretation and physical applications. Specific examples of geometric algebras applied in physics include the algebra of physical space, the spacetime algebra, and the conformal geometric algebra. Geometric calculus, an extension of GA that incorporates differentiation and integration, can be used to formulate other theories such as complex analysis, differential geometry, e.g. by using the Clifford algebra instead of differential forms. Geometric algebra has been advocated, most notably by David Hestenes and Chris Doran, as the preferred mathematical framework for physics. Proponents claim that it provides compact and intuitive descriptions in many areas including classical and quantum mechanics, electromagnetic theory and relativity. GA has also found use as a computational tool in computer graphics and robotics.

The geometric product was first briefly mentioned by Hermann Grassmann, who was chiefly interested in developing the closely related exterior algebra, which is the geometric algebra of the trivial quadratic form. In 1878, William Kingdon Clifford greatly expanded on Grassmann's work to form what are now usually called Clifford algebras in his honor (although Clifford himself chose to call them "geometric algebras"). For several decades, geometric algebras went somewhat ignored, greatly eclipsed by the vector calculus then newly developed to describe electromagnetism. The term "geometric algebra" was repopularized by Hestenes in the 1960s, who advocated its importance to relativistic physics.

DEFINITION AND NOTATION

Given a finite-dimensional real quadratic space V with a symmetric bilinear form (e.g. the Euclidean or Lorentzian metric) $g : V \times V \to R$, the geometric algebra for this quadratic space is the Clifford algebra $C\ell(V, g)$.

The algebra product is called the geometric product. It is standard to denote the geometric product by juxtaposition (i.e., suppressing any explicit multiplication symbol). The above definition of the geometric algebra is abstract, so we summarize the properties of the geometric product by the following set of axioms. The geometric product has the following properties:

$1a = a1$, where 1 is the unit and a any element of the algebra (existence of the unit)

$a(bc) = (ab)c$, where a, b and c are any elements of the algebra (associativity)

$a(b + c) = ab + ac$ and $(b + c)a = ba + ca$, where a, b and c are any elements of the algebra (distributivity)

$a^2 = g(a, a)1$, where a is any element of the subspace V of the algebra.

Note that in the final property above, the square of a vector need not be nonnegative if g is not positive-definite. An important property of the geometric product is the existence of elements with multiplicative inverse, also known as units. If $a^2 \neq 0$ for some vector a, then a^{-1} exists and is equal to a/a^2. Not every nonzero element of the algebra is necessarily a unit. For example, if u is a vector in V such that $u^2 = 1$, the elements $1 \pm u$ are zero divisors and thus have no inverse: $(1 - u)(1 + u) = 1 - uu = 1 - 1 = 0$. There may also exist nontrivial idempotent elements such as $(1 + u)/2$.

Inner and Exterior Product of Vectors

For vectors a and b, we may write the geometric product of any two vectors a and b as the sum of a symmetric product and an antisymmetric product:

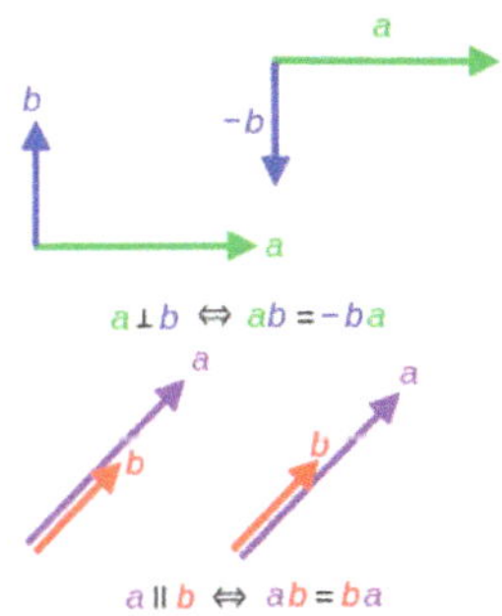

Given two vectors a and b, if the geometric product ab isanticommutative; they are perpendicular (top) because a ∧ b = −(b ∧ a) and a · b = 0, if it is commutative; they are parallel (bottom) because a ∧ b = 0 and a · b = b · a.

$$ab = \frac{1}{2}(ab + ba) + \frac{1}{2}(ab - ba)$$

Thus we can define the inner product of vectors as

$a \cdot b := g(a, b),$ so that the symmetric product can be written as

$$\frac{1}{2}(ab + ba) = \frac{1}{2}((a+b)^2 - a^2 - b^2) = a \cdot b,$$

Conversely, g is completely determined by the algebra. The antisymmetric part is the exterior product (also called the outer product) of the two vectors, the product of the contained exterior algebra:

$$a \wedge b := \frac{1}{2}(ab - ba) = -(b \wedge a)$$

The inner and exterior products are associated with familiar concepts from standard vector algebra. Pictorially, a and b are parallel if their geometric product is equal to their inner product, whereas a and b are perpendicular if their geometric product is equal to their exterior product. In a geometric algebra for which the square of any nonzero vector is positive, the inner product of two vectors can be identified with the dot product of standard vector algebra. The exterior product of two vectors can be identified with the signed area enclosed by a parallelogram the sides of which are the vectors. The cross product of two vectors in 3 dimensions with positive-definite quadratic form is closely related to their exterior product.

Most instances of geometric algebras of interest have a nondegenerate quadratic form. If the quadratic form is fully degenerate, the inner product of any two vectors is always zero, and the geometric algebra is then simply an exterior algebra. Unless otherwise stated, this article will treat only nondegenerate geometric algebras.

The exterior product is naturally extended as a completely antisymmetric, multilinear operator between any number of vectors:

$$a_1 \wedge a_2 \wedge \cdots \wedge a_r = \frac{1}{r!} \sum_{\sigma \in \mathfrak{S}_r} \operatorname{sgn}(\sigma) a_{\sigma(1)} a_{\sigma(2)} \cdots a_{\sigma(r)},$$

where the sum is over all permutations of the indices, with $\mathrm{sgn}(\sigma)$ the sign of the permutation.

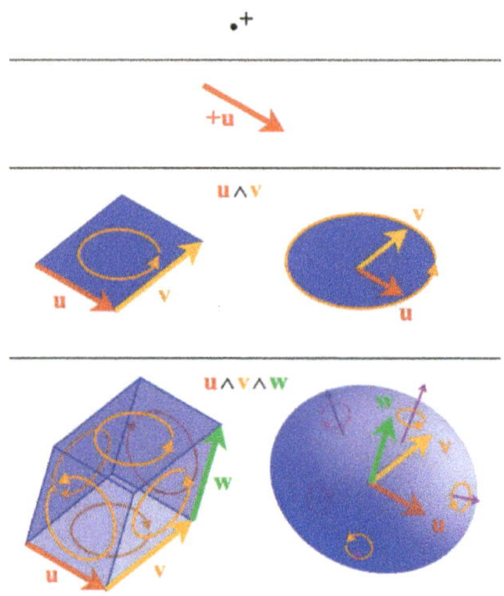

Orientation defined by an ordered set of vectors.

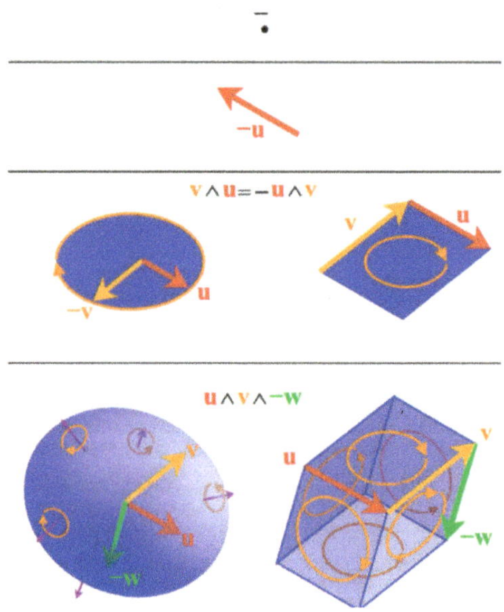

Reversed orientation corresponds to negating the exterior product.

Blades, grades, and canonical basis

A multivector that is the exterior product of r independent vectors ($r \leq n$) is called a blade, and the blade is said to be a multivector of grade r. From the axioms, with closure, every multivector of the geometric algebra is a sum of blades.

Consider a set of r independent vectors $\{a_1, \ldots, a_r\}$ spanning an r-dimensional subspace of the vector space. With these, we can define a real symmetric matrix

$$[\mathbf{A}]_{ij} = a_i \cdot a_j$$

By the spectral theorem, A can be diagonalized to diagonal matrix D by an orthogonal matrix O via

$$\sum_{k,l} [\mathbf{O}]_{ik} [\mathbf{A}]_{kl} [\mathbf{O}^{\mathrm{T}}]_{lj} = \sum_{k,l} [\mathbf{O}]_{ik} [\mathbf{O}]_{jl} [\mathbf{A}]_{kl} = [\mathbf{D}]_{ij}$$

Define a new set of vectors $\{e_1, \ldots, e_r\}$, known as orthogonal basis vectors, to be those transformed by the orthogonal matrix:

$$e_i = \sum_j [\mathbf{O}]_{ij} a_j$$

Since orthogonal transformations preserve inner products, it follows that $e_i \cdot e_j = [\mathbf{D}]_{ij}$ and thus the $\{e_1, \ldots, e_r\}$ are perpendicular. In other words, the geometric product of two distinct vectors $e_i \neq e_j$ is completely specified by their exterior product, or more generally

$$\begin{aligned} e_1 e_2 \cdots e_r &= e_1 \wedge e_2 \wedge \cdots \wedge e_r \\ &= \left(\sum_j [\mathbf{O}]_{1j} a_j\right) \wedge \left(\sum_j [\mathbf{O}]_{2j} a_j\right) \wedge \cdots \wedge \left(\sum_j [\mathbf{O}]_{rj} a_j\right) \\ &= \det[\mathbf{O}] a_1 \wedge a_2 \wedge \cdots \wedge a_r \end{aligned}$$

Therefore, every blade of grade r can be written as a geometric product of r vectors. More generally, if a degenerate geometric algebra is allowed, then the orthogonal matrix is replaced by a block matrix that is orthogonal in the nondegenerate block, and the diagonal matrix has zero-valued entries along the degenerate dimensions. If the new vectors of the nondegenerate subspace are normalized according to

$$\hat{e}_i = \frac{1}{\sqrt{|e_i \cdot e_i|}} e_i,$$

then these normalized vectors must square to +1 or -1. By Sylvester's law of inertia, the total number of +1s and the total number of -1s along the diagonal matrix is invariant. By extension, the total number p of these vectors that square to +1 and the total number q that square to -1 is invariant. (If the degenerate case is allowed, then the total number of basis vectors that square to zero is also invariant.) We denote this algebra $\mathcal{G}$ (p,q). For example, $\mathcal{G}(3,0)$ models 3-dimensional Euclidean space, $\mathcal{G}(1,3)$ relativistic spacetime and $\mathcal{G}$ (4,1) a conformal geometric algebra of a 3-dimensional space.

$$\{1, e_1, e_2, e_3, e_1e_2, e_1e_3, e_2e_3, e_1e_2e_3\}$$

A basis formed this way is called a canonical basis for the geometric algebra, and any other orthogonal basis for V will produce another canonical basis. Each canonical basis consists of 2^n elements. Every multivector of the geometric algebra can be expressed as a linear combination of the canonical basis elements. If the canonical basis elements are $\{B_i \mid i \in S\}$ with S being an index set, then the geometric product of any two multivectors is

$$\left(\sum_i \alpha_i B_i\right)\left(\sum_j \beta_j B_j\right) = \sum_{i,j} \alpha_i \beta_j B_i B_j.$$

Grade Projection

Using a canonical basis, a graded vector space structure can be established. Elements of the geometric algebra that are simply scalar multiples of 1 are grade-0 blades and are called scalars. Nonzero multivectors that are in the span of $\{e_1, \ldots, e_n\}$ are grade-1 blades and are the ordinary vectors. Multivectors in the span of $\{e_i e_j \mid 1 \leq i < j \leq n\}$ are grade-2 blades and are the bivectors. This terminology continues through to the last grade of n-vectors. Alternatively, grade-n blades are called pseudoscalars, grade-n−1 blades pseudovectors, etc. Many of the elements of the algebra are not graded by this scheme since they are sums of elements of differing grade. Such elements are said to be of mixed grade. The grading of multivectors is independent of the orthogonal basis chosen originally.

A multivector A may be decomposed with the grade-projection operator $(A)_r$, which outputs the grade-r portion of A. As a result:

$$A = \sum_{r=0}^{n} \langle A \rangle_r$$

As an example, the geometric product of two vectors $ab = a \cdot b + a \wedge b = \langle ab \rangle_0 + \langle ab \rangle_2$ since $\langle ab \rangle_0 = a \cdot b$ and $\langle ab \rangle_2 = a \wedge b$ and $\langle ab \rangle_i = 0$, for i other than 0 and 2.

This makes the algebra a $\mathbf{Z}_2$-graded algebra or superalgebra with the geometric product. Since the geometric product of two even multivectors is an even multivector, they define an even subalgebra. The even subalgebra of an n-dimensional geometric algebra is isomorphic to a full geometric algebra of (n−1) dimensions. Examples include $\mathcal{G}^+(2,0) \cong \mathcal{G}(0,1)$ and $\mathcal{G}^+(1,3) \cong \mathcal{G}(3,0)$

Representation of Subspaces

Geometric algebra represents subspaces of V as multivectors, and so they coexist in the same algebra with vectors from V. A k-dimensional subspace W of V is represented by taking an orthogonal basis $\{b_1, b_2, \ldots, b_k\}$ and using the geometric product to form the blade $D = b_1 b_2 \cdots b_k$. There are multiple blades representing W; all those representing W are scalar multiples of D. These blades can be separated into two sets: positive multiples of D and negative multiples of D. The positive multiples of D are said to have the same orientation as D, and the negative multiples the opposite orientation.

Blades are important since geometric operations such as projections, rotations and reflections depend on the factorability via the exterior product that (the restricted class of) n-blades provide but that (the generalized class of) grade-n multivectors do not when $n \geq 4$.

Unit Pseudoscalars

Unit pseudoscalars are blades that play important roles in GA. A unit pseudoscalar for a non-degenerate subspace W of V is a blade that is the product of the members of an orthonormal basis for W. It can be shown that if I and I′ are both unit pseudoscalars for W, then $I = \pm I'$ and $I^2 = \pm 1$.

Suppose the geometric algebra $\mathcal{G}(n,0)$ with the familiar positive definite inner product on $\mathbf{R}^n$ is formed. Given a plane (2-dimensional subspace) of $\mathbf{R}^n$, one can find an orthonormal basis $\{b_1, b_2\}$ spanning the plane, and thus find a unit pseudoscalar $I = b_1 b_2$ representing this plane. The geometric product of any two vectors in the span of b_1 and b_2 lies in $\{\alpha_0 + \alpha_1 I \mid \alpha_i \in \mathbf{R}\}$, that is, it is the sum of a 0-vector and a 2-vector.

By the properties of the geometric product, $I^2 = b_1 b_2 b_1 b_2 = -b_1 b_2 b_2 b_1 = -1$. The resemblance to the imaginary unit is not accidental: the subspace $\{\alpha_0 + \alpha_1 I \mid \alpha_i \in \mathbf{R}\}$ is $\mathbf{R}$ -algebra isomorphic to the complex numbers. In this way, a copy of the complex numbers is embedded in the geometric algebra for each 2-dimensional subspace of V on which the quadratic form is definite.

It is sometimes possible to identify the presence of an imaginary unit in a physical equation. Such units arise from one of the many quantities in the real algebra that square to −1, and these have geometric significance because of the properties of the algebra and the interaction of its various subspaces.

In $\mathcal{G}(3,0)$, , an exceptional case occurs. Given a canonical basis consisting of orthonormal vectors e_i of V, , the set of all 2-vectors is generated by $\{e_3 e_2, e_1 e_3, e_2 e_1\}$.

Labelling these i, j and k (momentarily deviating from our uppercase convention), the subspace generated by 0-vectors and 2-vectors is exactly $\{\alpha_0 + i\alpha_1 + j\alpha_2 + k\alpha_3 \mid \alpha_i \in \mathbf{R}\}$. This set is seen to be a subalgebra, and furthermore is R-algebra isomorphic to the quaternions, another important algebraic system.

Dual Basis

Let $\{e_i\}$ be a basis of V, i.e. a set of n linearly independent vectors that span the n-dimensional vector space V. The basis that is dual to $\{e_i\}$ is the set of elements of the dual vector space V* that forms a biorthogonal system with this basis, thus being the elements denoted $\{e_i\}$ satisfying

$$e^i \cdot e_j = \delta^i{}_j,$$

where δ is the Kronecker delta.

Given a nondegenerate quadratic form on V, V* becomes naturally identified with V, and the dual basis may be regarded as elements of V, but are not in general the same set as the original basis.

Given further a GA of V, let $\epsilon = e_1 \wedge \cdots \wedge e_n$ be the pseudoscalar (which does not necessarily square to ± 1) formed from the basis $\{e_i\}$. The dual basis vectors may be constructed as

$$e^i = (-1)^{i-1}(e_1 \wedge \cdots \wedge \check{e}_i \wedge \cdots \wedge e_n)\epsilon^{-1},$$

where the $\check{e}_i$ denotes that the ith basis vector is omitted from the product.

Extensions of the Inner and Exterior Products

It is common practice to extend the exterior product on vectors to the entire algebra. This may be done through the use of the grade projection operator:

$$C \wedge D := \sum_{r,s} \langle \langle C \rangle_r \langle D \rangle_s \rangle_{r+s} \text{ (the exterior product)}$$

This generalization is consistent with the above definition involving antisymmetrization. Another generalization related to the exterior product is the commutator product:

$$C \times D := \tfrac{1}{2}(CD - DC)$$

The regressive product is the dual of the exterior product:

$C \vee D := ((CI^{-1}) \wedge (DI^{-1}))I$ with I the unit pseudoscalar of the algebra. The regressive product, like the exterior product, is associative.

The inner product on vectors can also be generalized, but in more than one non-equivalent way. The paper (Dorst 2002) gives a full treatment of several different inner products developed for geometric algebras and their interrelationships, and the notation is taken from there. Many authors use the same symbol as for the inner product of vectors for their chosen extension (e.g. Hestenes and Perwass). No consistent notation has emerged.

Among these several different generalizations of the inner product on vectors are:

$$C \rfloor D := \sum_{r,s} \langle \langle C \rangle_r \langle D \rangle_s \rangle_{s-r} \text{ (the \textit{left contraction})}$$

$$C \lfloor D := \sum_{r,s} \langle \langle C \rangle_r \langle D \rangle_s \rangle_{r-s} \text{ (the \textit{right contraction})}$$

$$C * D := \sum_{r,s} \langle \langle C \rangle_r \langle D \rangle_s \rangle_0 \text{ (the \textit{scalar product})}$$

$$C \bullet D := \sum_{r,s} \langle \langle C \rangle_r \langle D \rangle_s \rangle_{|s-r|} \text{ (the "(fat) dot" product)}$$

Dorst (2002) makes an argument for the use of contractions in preference to Hestenes's inner product; they are algebraically more regular and have cleaner geometric interpretations. A number of identities incorporating the contractions are valid without restriction of their inputs. For example,

$$C \rfloor D = (C \wedge (DI^{-1}))I$$
$$C \lfloor D = I((I^{-1}C) \wedge D)$$
$$A \rfloor (B \rfloor C) = (A \wedge B) \rfloor C$$
$$A \bullet B = A \rfloor B + A \lfloor B - A * B.$$

Benefits of using the left contraction as an extension of the inner product on vectors include that the identity $ab = a \cdot b + a \wedge b$ is extended to $aB = a \rfloor B + a \wedge B$ for any vector a for any vector B, and that the projection operation $\mathcal{P}_b(a) = (a \cdot b^{-1})b$ is extended to $\mathcal{P}_B(A) = (A \rfloor B^{-1}) \rfloor B$ for any blade B and any multivector A (with a minor modification to accommodate null B, given below).

Terminology Specific to Geometric Algebra

Some terms are used in geometric algebra with a meaning that differs from the use of those terms in other fields of mathematics. Some of these are listed here:

Vector

In GA this refers to an element of the 1-vector subspace unless otherwise clear from the context, despite the entire algebra forming a vector space.

Grade

In GA this refers to a grading as an algebra under the exterior product (an N-grading), and not under the geometric product (which produces a Z_2^n-grading).

Outer product

When used in GA, this is a synonym for the exterior product. It is not the outer product of linear algebra.

Inner product

When used in GA, this generally refers to the symmetric bilinear form on the 1-vector subspace, not the inner product on a normed vector space. Some authors may extend its meaning to the entire algebra, but there is little consensus on this. Even in texts on geometric algebras, the term is not universally used.

Versor

In GA this refers to an element of the algebra that can be constructed as the geometric product of any number of non-null vectors; an element of the Clifford group. Elsewhere, the term may refer to a unit quaternion, analogous to a rotor in GA.

Outermorphism

This term is used only in GA, and refers to a linear map on the vector subspace, extended to apply to the entire algebra by defining it as preserving the exterior product.

GEOMETRIC INTERPRETATION

Projection and Rejection

For any vector a and any invertible vector m,

$$a = amm^{-1} = (a \cdot m + a \wedge m)m^{-1} = a_{\|m} + a_{\perp m},$$

where the projection of a onto m (or the parallel part) is

$$a_{\|m} = (a \cdot m)m^{-1}$$

and the rejection of a onto m (or the perpendicular part) is $a_{\perp m} = a - a_{\| m} = (a \wedge m)m^{-1}$.

Using the concept of a k-blade B as representing a subspace of V and every multivector ultimately being expressed in terms of vectors, this generalizes to projection of a general multivector onto any invertible k-blade B as

$$\mathcal{P}_B(A) = (A \rfloor B^{-1}) \rfloor B,$$

with the rejection being defined as

$$\mathcal{P}_B^{\perp}(A) = A - \mathcal{P}_B(A).$$

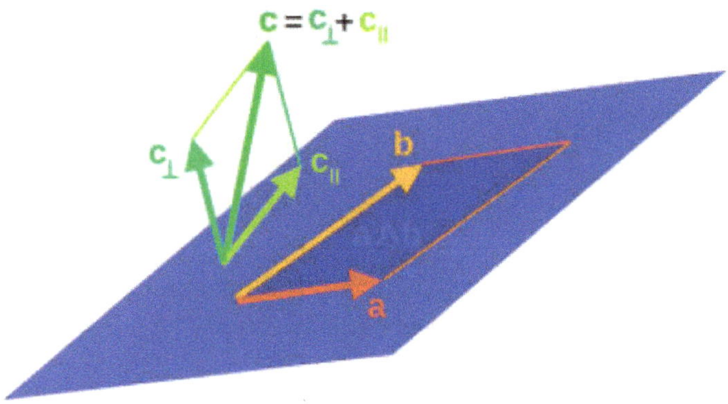

In 3d space, a bivector **a∧b** defines a 2d plane subspace (light blue, extends infinitely in indicated directions). Any vector **c** in the 3-space can be projected onto and rejected normal to the plane, shown respectively by $\mathbf{c}_{\perp}$ and $\mathbf{c}_{\|}$.

The projection and rejection generalize to null blades B by replacing the inverse B^{-1} with the pseudoinverse B^{+} with respect to the contractive product. The outcome of the projection coincides in both cases for non-null blades. For null blades B, the definition of the projection given here with the first contraction rather than the second being onto the pseudoinverse should be used, as only then is the result necessarily in the subspace represented by B. The projection generalizes through linearity to general multivectors A. The projection is not linear in B and does not generalize to objects B that are not blades.

Reflections

The definition of a reflection occurs in two forms in the literature. Several authors work with reflection on a vector (negating all vector components except that parallel to the specifying vector), while others work with reflection along a vector (negating only the component parallel to the specifying vector, or reflection in the hypersurface orthogonal to that vector). Either may be used to build general versor operations, but the former has the advantage that it extends to the algebra in a simpler and algebraically more regular fashion.

Reflection *on* a Vector

The result (c′) of reflecting a vector c on another vector n is to negate the rejection of c. It is akin to reflecting the vector c through the origin, except that the projection of c onto n is not changed. Such an operation is described by

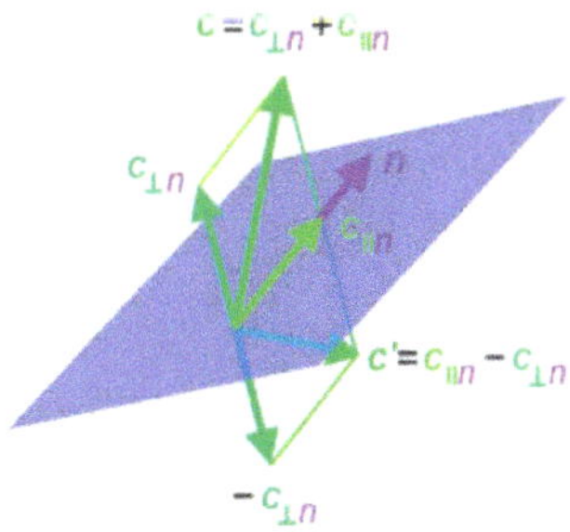

Reflection of vector c on a vector n. The rejection of c on n is negated.

$$c \mapsto c' = ncn^{-1}.$$

Repeating this operation with several different vectors n results in a general versor operation (including both rotations and reflections) of a general multivector A being expressed as

$$A \mapsto NAN^{-1},$$

where N is the geometric product of several vectors. This allows a general definition of any versor N (including both reflections and rotors) as an object that can be expressed as a geometric product of any number of non-null 1-vectors. Such a versor can be applied in a uniform sandwich product as above irrespective of whether it is of even (a proper rotation) or odd grade (an improper rotation i.e. general reflection). The set of all versors with the geometric product as the group operation constitutes the Clifford group of the Clifford algebra $C\ell_{p,q}(\mathbf{R})$.

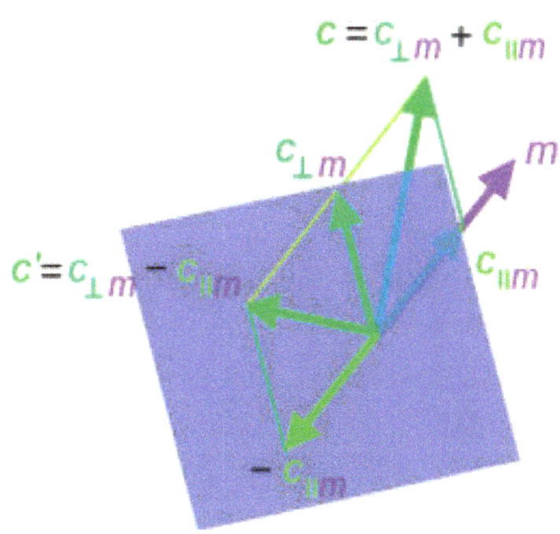

Reflection of vector c along a vector m. Only the component of c parallel to m is negated.

Reflection along a Vector

The reflection (c') of a vector c along a vector m, or equivalently in the hyperplane orthogonal to m, is the same as negating the component of a vector parallel to m. The result of the reflection will be

$$c' = -c_{\|m} + c_{\perp m} = -(c \cdot m)m^{-1} + (c \wedge m)m^{-1} = (-m \cdot c - m \wedge c)m^{-1} = -mcm^{-1}$$

This is not the most general operation that may be regarded as a reflection when the dimension n ≥ 4. A general reflection may be expressed as the composite of any odd number of single-axis reflections. Thus, a general reflection (a′) of a vector a may be written

$$a \mapsto a' = -MaM^{-1}$$

where

$$M = pq \cdots r \text{ and} M^{-1} = (pq \cdots r)^{-1} = r^{-1} \cdots q^{-1}p^{-1}.$$

If we define the reflection along a non-null vector m of the product of vectors as the reflection of every vector in the product along the same vector, we get for any product of an odd number of vectors that, by way of example,

$$(abc)' = a'b'c' = (-mam^{-1})(-mbm^{-1})(-mcm^{-1}) = -ma(m^{-1}m)b(m^{-1}m)cm^{-1} = -mabcm^{-1}$$

and for the product of an even number of vectors that

$$(abcd)' = a'b'c'd' = (-mam^{-1})(-mbm^{-1})(-mcm^{-1})(-mdm^{-1}) = mabcdm^{-1}.$$

Using the concept of every multivector ultimately being expressed in terms of vectors, the reflection of a general multivector A using any reflection versor M may be written

$$A \mapsto M\alpha(A)M^{-1},$$

where α is the automorphism of reflection through the origin of the vector space (v ↦ −v) extended through linearity to the whole algebra.

Hypervolume of an n-Parallelotope Spanned by n Vectors

For vectors a and b spanning a parallelogram we have

$$a \wedge b = ((a \wedge b)b^{-1})b = a_{\perp b}b$$

with the result that $a \wedge b$ is linear in the product of the "altitude" and the "base" of the parallelogram, that is, its area.

Similar interpretations are true for any number of vectors spanning an n-dimensional parallelotope; the exterior product of vectors a_1, a_2, ..., a_n, that is $\bigwedge_{i=1}^{n} a_i$, has a magnitude equal to the volume of the n-parallelotope. An n-vector doesn't necessarily have a shape of a parallelotope – this is a convenient visualization. It could be any shape, although the volume equals that of the parallelotope.

Rotations

If we have a product of vectors $R = a_1 a_2 \cdots a_r$ then we denote the reverse as

$$R^\dagger = (a_1 a_2 \cdots a_r)^\dagger = a_r \cdots a_2 a_1.$$

As an example, assume that $R = ab$ we get

$$RR^\dagger = abba = ab^2 a = a^2 b^2 = ba^2 b = baab = R^\dagger R.$$

Scaling R so that $RR^\dagger = 1$ then

$$(RvR^\dagger)^2 = Rv^2 R^\dagger = v^2 RR^\dagger = v^2$$

so $RvR^\dagger$ leaves the length of v unchanged. We can also show that

$$(Rv_1 R^\dagger) \cdot (Rv_2 R^\dagger) = v_1 \cdot v_2$$

so the transformation $RvR^\dagger$ preserves both length and angle. It therefore can be identified as a rotation or rotoreflection; R is called a rotor if it is a proper rotation (as it is if it can be expressed as a product of an even number of vectors) and is an instance of what is known in GA as a versor (presumably for historical reasons).

There is a general method for rotating a vector involving the formation of a multivector of the form $R = e^{-\frac{B\theta}{2}}$ that produces a rotation θ in the plane and with the orientation defined by a 2-blade B.

Rotors are a generalization of quaternions to n-D spaces.

For more about reflections, rotations and "sandwiching" products like $RvR^\dagger$ see Plane of rotation.

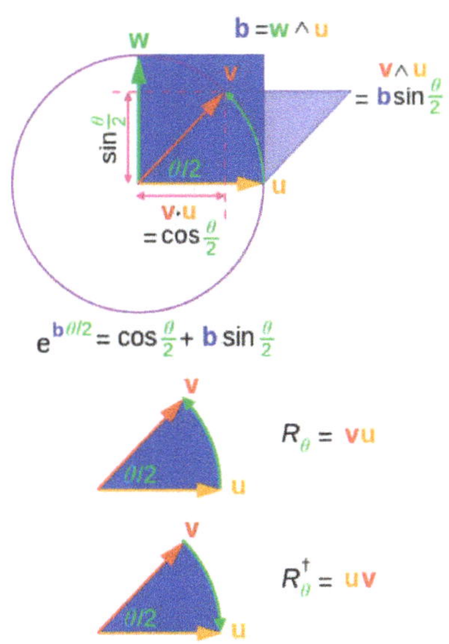

A rotor that rotates vectors in a plane rotates vectors through angle θ, that is $x \mapsto R_\theta x R_\theta^\dagger$ is a rotation of x through angle θ. The angle between u and v is θ/2. Similar interpretations are valid for a general multivector X instead of the vector x.

Linear Functions

An important class of functions of multivectors are the linear functions mapping multivectors to multivectors. The geometric algebra of an n-dimensional vector space is spanned by a basis of 2^n elements. If a multivector is represented by a $2^n \times 1$ real column matrix in terms of a basis, then all linear transformations of the multivector can be expressed as the matrix multiplication by a $2^n \times 2^n$ real matrix. However, such a general linear transformation allows arbitrary exchanges among grades, such as a "rotation" of a scalar into a vector, which has no clear geometric interpretation.

A general linear transformation from vectors to vectors is of interest. Thought there is no unique generalization of these transformations to the entire geometric algebra without further restriction, a natural choice is that of the outermorphism of the linear transformation. It extends the transformation through the expression of a multivector as a sum-of-products-of-vectors within the algebra and applies a given linear map to each vector in the expression. If f is a linear function that maps vectors to vectors, then its outermorphism is the function that obeys the rule

$$\underline{f}(a_1 \wedge a_2 \wedge \cdots \wedge a_r) = f(a_1) \wedge f(a_2) \wedge \cdots \wedge f(a_r)$$

for a blade, extended to the whole algebra through linearity.

In the cases of reflections and rotations, their outermorphisms have a particularly simple algebraic form. Specifically, a mapping of vectors of the form

$$V \to V : a \mapsto RaR^{-1},$$

extends to the outermorphism

$$\mathcal{G}(V) \to \mathcal{G}(V) : A \mapsto RAR^{-1}.$$

EXAMPLES AND APPLICATIONS

Intersection of a Line and a Plane

We may define the line parametrically by $p = t + \alpha v$ where p and t are position vectors for points T and P and v is the direction vector for the line.

Then

$$B \wedge (p - q) = 0 \text{ and } B \wedge (t + \alpha v - q) = 0$$

so

$$\alpha = \frac{B \wedge (q - t)}{B \wedge v}$$

and

$$p = t + \left(\frac{B \wedge (q - t)}{B \wedge v}\right) v.$$

Rotating Systems

The mathematical description of rotational forces such as torque and angular momentum often makes use of the cross product of vector calculus in three dimensions with a convention of orientation (handedness).

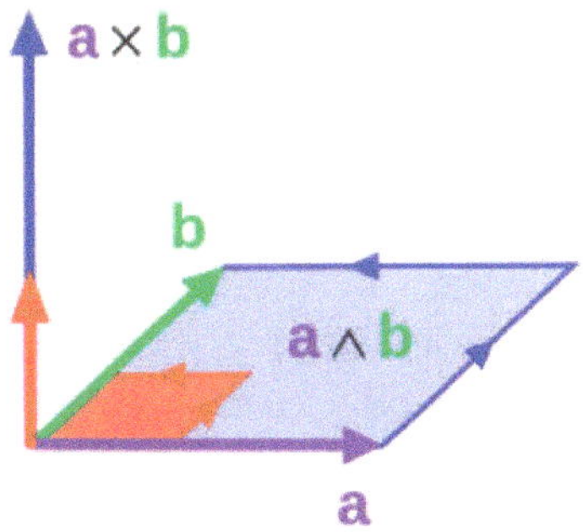

The cross product in relation to the exterior product. In red are the unit normal vector, and the "parallel" unit bivector.

The cross product can be viewed in terms of the exterior product allowing a more natural geometric interpretation of the cross product as a bivector using the dual relationship

$$a \times b = -I(a \wedge b).$$

For example, torque is generally defined as the magnitude of the perpendicular force component times distance, or work per unit angle.

Suppose a circular path in an arbitrary plane containing orthonormal vectors $\hat{u}$ and $\hat{v}$ parameterized by angle.

$$\mathbf{r} = r(\hat{u}\cos\theta + \hat{v}\sin\theta) = r\hat{u}(\cos\theta + \hat{u}\hat{v}\sin\theta)$$

By designating the unit bivector of this plane as the imaginary number

$$i = \hat{u}\hat{v} = \hat{u} \wedge \hat{v}$$

$$i^2 = -1$$

this path vector can be conveniently written in complex exponential form

$$\mathbf{r} = r\hat{u}e^{i\theta}$$

and the derivative with respect to angle is

$$\frac{d\mathbf{r}}{d\theta} = r\hat{u}ie^{i\theta} = \mathbf{r}i$$

So the torque, the rate of change of work W, due to a force F, is

$$\tau = \frac{dW}{d\theta} = F \cdot \frac{dr}{d\theta} = F \cdot (\mathbf{r}i)$$

Unlike the cross product description of torque, $\tau = \mathbf{r} \times F$, the geometric algebra description does not introduce a vector in the normal direction; a vector that does not exist in two and that is not unique in greater than three dimensions. The unit bivector describes the plane and the orientation of the rotation, and the sense of the rotation is relative to the angle between the vectors $\hat{u}$ and $\hat{v}$

Electrodynamics and Special Relativity

In physics, the main applications are the geometric algebra of Minkowski 3+1 spacetime, $C\ell_{1,3}$, called spacetime algebra (STA). or less commonly, $C\ell_3$, called the algebra of physical space (APS) where $C\ell_3$ is isomorphic to the even subalgebra of the 3+1 Clifford algebra, $C\ell 0_3$,1.

While in STA points of spacetime are represented simply by vectors, in APS, points of (3+1)-dimensional spacetime are instead represented by paravectors: a 3-dimensional vector (space) plus a 1-dimensional scalar (time).

In spacetime algebra the electromagnetic field tensor has a bivector representation $F = (E + icB)e_0$ Here, the $i = \gamma_0\gamma_1\gamma_2\gamma_3$ is the unit pseudoscalar (or four-dimensional volume element), γ_0 is the unit vector in time direction, and E and B are the classic electric and magnetic field vectors (with a zero time component). Using the four-current J, Maxwell's equations then become

Formulation	Homogeneous equations	Non-homogeneous equations
Fields	$DF = \mu_0 J$	
	$D \wedge F = 0$	$D \cdot F = \mu_0 J$
Potentials (any gauge)	$F = D \wedge A$	$D \cdot D \wedge A = \mu_0 J$
Potentials (Lorenz gauge)	$F = DA$ $D \cdot A = 0$	$D^2 A = \mu_0 J$

In geometric calculus, juxtapositioning of vectors such as in DF indicate the geometric product and can be decomposed into parts as $DF = D \cdot F + D \wedge F$. Here D is the covector derivative in any spacetime and reduces to to ∇ in flat spacetime. Where ∇ plays a role in Minkowski 4 -spacetime which is synonymous to the role of ∇ in Euclidean 3 -space and is related to the D'Alembertian by $\Box = \nabla^2$. Indeed, given an observer represented by a future pointing timelike vector γ_0 we have

$$\gamma_0 \cdot \nabla = \frac{1}{c}\frac{\partial}{\partial t}$$

$$\gamma_0 \wedge \nabla = \nabla$$

Boosts in this Lorentzian metric space have the same expression e^{β} as rotation in Euclidean space, where β is the bivector generated by the time and the space directions involved, whereas in the Euclidean case it is the bivector generated by the two space directions, strengthening the "analogy" to almost identity.

Relationship with Other Formalisms

$\mathcal{G}(3,0)$ may be directly compared to vector algebra.

The even subalgebra of $\mathcal{G}(2,0)$ is isomorphic to the complex numbers, as may be seen by writing a vector P n terms of its components in an orthonormal basis and left multiplying by the basis vector e1, yielding

$$Z = e_1 P = e_1(xe_1 + ye_2) = x(1) + y(e_1 e_2),$$

where we identify $i \mapsto e_1 e_2$ since

$$(e_1 e_2)^2 = e_1 e_2 e_1 e_2 = -e_1 e_1 e_2 e_2 = -1.$$

Every associative algebra has a matrix representation; the Pauli matrices are a representation of $\mathcal{G}(3,0)$ and the Dirac matrices are a representation of $\mathcal{G}(1,3)$ showing the equivalence with matrix representations used by physicists.

Geometric Calculus

Geometric calculus extends the formalism to include differentiation and integration including differential geometry and differential forms.

Essentially, the vector derivative is defined so that the GA version of Green's theorem is true,

$$\int_A dA\, \nabla f = \oint_{\partial A} dx\, f$$

and then one can write

$$\nabla f = \nabla \cdot f + \nabla \wedge f$$

as a geometric product, effectively generalizing Stokes' theorem (including the differential form version of it).

In $1D$ when A is a curve with endpoints a and b, then

$$\int_A dA\, \nabla f = \oint_{\partial A} dx\, f$$

reduces to

$$\int_a^b dx\, \nabla f = \int_a^b dx \cdot \nabla f = \int_a^b df = f(b) - f(a)$$

or the fundamental theorem of integral calculus.

Also developed are the concept of vector manifold and geometric integration theory (which generalizes Cartan's differential forms).

Conformal Geometric Algebra (CGA)

A compact description of the current state of the art is provided by Bayro-Corrochano and Scheuermann (2010), which also includes further references, in particular to Dorst et al (2007). Other useful references are Li (2008) and Bayro (2010).

Working within GA, Euclidean space $\mathcal{E}^3$ is embedded projectively in the CGA $\mathcal{G}^{4,1}$ via the identification of Euclidean points with 1D subspaces in the 4D null cone of the 5D CGA vector subspace, and adding a point at infinity. This allows all conformal transformations to be done as rotations and reflections and is covariant, extending incidence relations of projective geometry to circles and spheres.

Specifically, we add orthogonal basis vectors e_+ and e_- such that $e_+{}^2 = +1$ and $e_-{}^2 = -1$ to the basis of $\mathcal{G}(3,0)$ and identify null vectors

$n_\infty = e_- + e_+$ as an ideal point (point at infinity) (see Compactification) and

$n_o = \frac{1}{2}(e_- - e_+)$ as the point at the origin, giving $n_\infty \cdot n_o = -1$.

This procedure has some similarities to the procedure for working with homogeneous coordinates in projective geometry and in this case allows the modeling of Euclidean transformations as orthogonal transformations.

A fast changing and fluid area of GA, CGA is also being investigated for applications to relativistic physics.

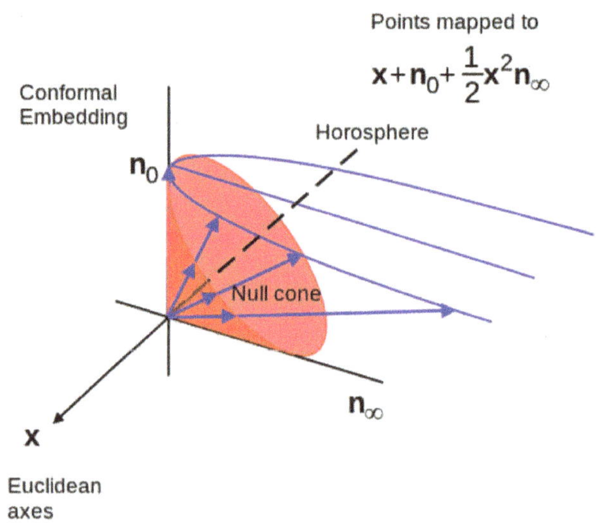

Before the 20th century

Although the connection of geometry with algebra dates as far back at least to Euclid's Elements in the 3rd century B.C. (see Greek geometric algebra), GA in the sense used in this article was not developed until 1844, when it was used in a systematic way to describe the geometrical properties and transformations of a space. In that year, Hermann

Grassmann introduced the idea of a geometrical algebra in full generality as a certain calculus (analogous to the propositional calculus) that encoded all of the geometrical information of a space. Grassmann's algebraic system could be applied to a number of different kinds of spaces, the chief among them being Euclidean space, affine space, and projective space. Following Grassmann, in 1878 William Kingdon Clifford examined Grassmann's algebraic system alongside the quaternions of William Rowan Hamilton in (Clifford 1878). From his point of view, the quaternions described certain transformations (which he called rotors), whereas Grassmann's algebra described certain properties (or Strecken such as length, area, and volume). His contribution was to define a new product — the **geometric product** — on an existing Grassmann algebra, which realized the quaternions as living within that algebra. Subsequently Rudolf Lipschitz in 1886 generalized Clifford's interpretation of the quaternions and applied them to the geometry of rotations in n dimensions. Later these developments would lead other 20th-century mathematicians to formalize and explore the properties of the Clifford algebra.

Nevertheless, another revolutionary development of the 19th-century would completely overshadow the geometric algebras: that of vector analysis, developed independently by Josiah Willard Gibbs and Oliver Heaviside. Vector analysis was motivated by James Clerk Maxwell's studies of electromagnetism, and specifically the need to express and manipulate conveniently certain differential equations. Vector analysis had a certain intuitive appeal compared to the rigors of the new algebras. Physicists and mathematicians alike readily adopted it as their geometrical toolkit of choice, particularly following the influential 1901 textbook Vector Analysis by Edwin Bidwell Wilson, following lectures of Gibbs.

In more detail, there have been three approaches to geometric algebra: quaternionic analysis, initiated by Hamilton in 1843 and geometrized as rotors by Clifford in 1878; geometric algebra, initiated by Grassmann in 1844; and vector analysis, developed out of quaternionic analysis in the late 19th century by Gibbs and Heaviside. The legacy of quaternionic analysis in vector analysis can be seen in the use of i, j, k to indicate the basis vectors of $\mathbf{R}^3$: it is being thought of as the purely imaginary quaternions. From the perspective of geometric algebra, quaternions can be identified as $C\ell^0_{3,0}(\mathbf{R})$, the even part of the Clifford algebra on Euclidean 3-space, which unifies the three approaches.

20th century and present

Progress on the study of Clifford algebras quietly advanced through the twentieth century, although largely due to the work of abstract algebraists such as Hermann Weyl and Claude Chevalley. The geometrical approach to geometric algebras has seen a number of 20th-century revivals. In mathematics, Emil Artin's Geometric Algebra discusses the algebra associated with each of a number of geometries, including affine geometry, projective geometry, symplectic geometry, and orthogonal geometry. In physics, geometric algebras have been revived as a "new" way to do classical mechanics and electromagnetism, together with more advanced topics such as quantum mechanics and gauge theory. David Hestenes reinterpreted the Pauli and Dirac matrices as vectors in ordinary space and spacetime, respectively, and has been a primary contemporary advocate for the use of geometric algebra.

In computer graphics and robotics, geometric algebras have been revived in order to efficiently represent rotations and other transformations. For applications of GA in robotics (screw theory, kinematics and dynamics using versors), computer vision, control and neural computing (geometric learning) see Bayro (2010).

Software

GA is a very application-oriented subject. There is a reasonably steep initial learning curve associated with it, but this can be eased somewhat by the use of applicable software.

The following is a list of freely available software that does not require ownership of commercial software or purchase of any commercial products for this purpose:

- GA Viewer Fontijne, Dorst, Bouma & Mann
- galgebra modules for sympy Alan Bromborsky

The link provides a manual, introduction to GA and sample material as well as the software.

- CLUViz Perwass

Software allowing script creation and including sample visualizations, manual and GA introduction.

- Gaigen Fontijne

For programmers, this is a code generator with support for C, C++, C# and Java.

- Cinderella Visualizations Hitzer and Dorst.
- Gaalop Standalone GUI-Application that uses the Open-Source Computer Algebra Software Maxima to break down CLUViz code into C/C++ or Java code.
- Gaalop Precompiler Precompiler based on Gaalop integrated with CMake.
- Gaalet, C++ Expression Template Library Seybold.
- Versor Colapinto, A lightweight templated C++ Library with an OpenGL interface for efficient geometric algebra programming in arbitrary metrics, including conformal
- Clifford Algebra with Mathematica clifford.m

REFERENCES

1. Baylis, W. E., ed. (1996), Clifford (Geometric) Algebra with Applications to Physics, Mathematics, and Engineering, Birkhäuser
2. Baylis, W. E. (2002), Electrodynamics: A Modern Geometric Approach (2 ed.), Birkhäuser, ISBN 978-0-8176-4025-5
3. Bourbaki, Nicolas (1980), Eléments de Mathématique. Algèbre, Ch. 9 “Algèbres de Clifford”: Hermann
4. Dorst, Leo (2002), The inner products of geometric algebra, Boston, MA: Birkhäuser Boston, p. 35–46

5. Hestenes, David (1999), New Foundations for Classical Mechanics (2 ed.), Springer Verlag, ISBN 978-0-7923-5302-7
6. Hestenes, David (1966). Space-time Algebra. New York: Gordon and Breach. ISBN 978-0-677-01390-9. OCLC 996371.
7. Lasenby, J.; Lasenby, A. N.; Doran, C. J. L. (2000), "A Unified Mathematical Language for Physics and Engineering in the 21st Century" (PDF), Philosophical Transactions of the Royal Society of London, pp. 1–18, 358 (A 358): 21, Bibcode:2000RSPTA.358...21L, doi:10.1098/rsta.2000.0517
8. Doran, Chris; Lasenby, Anthony (2003). Geometric algebra for physicists (PDF). Cambridge University Press. ISBN 978-0-521-71595-9.
9. Macdonald, Alan (2011). Linear and Geometric Algebra. Charleston: CreateSpace. ISBN 9781453854938. OCLC 704377582.
10. J Bain (2006). "Spacetime structuralism: §5 Manifolds vs. geometric algebra". In Dennis Dieks. The ontology of spacetime. Elsevier. p. 54 ff. ISBN 978-0-444-52768-4.
11. Bayro-Corrochano, Eduardo (2010). Geometric Computing for Wavelet Transforms, Robot Vision, Learning, Control and Action. Springer Verlag.
12. Vince, John A. (2008), Geometric Algebra for Computer Graphics, Springer, ISBN 978-1-84628-996-5
13. Christian Perwass (2009). Geometric Algebra with Applications in Engineering. Springer Science & Business Media. ISBN 978-3-540-89068-3.
14. Ron Goldman (2010). Rethinking Quaternions: Theory and Computation. Morgan & Claypool Publishers. Part III. Rethinking Quaternions and Clifford Algebras. ISBN 978-1-60845-420-4.
15. John Snygg (2011). A New Approach to Differential Geometry using Clifford's Geometric Algebra. Springer Science & Business Media. ISBN 978-0-8176-8282-8.
16. Dietmar Hildenbrand (2013). Foundations of Geometric Algebra Computing. Springer Science & Business Media. ISBN 978-3-642-31793-4.

CHAPTER 8

Inner Product Space

Chapter Contents

- **Introduction**
- **Examples**
- **Norms On Inner Product Spaces**
- **Orthonormal Sequences**
- **Generalizations**

An inner product is a generalization of the dot product. In a vector space, it is a way to multiply vectors together, with the result of this multiplication being a scalar.

More precisely, for a real vector space, an inner product $\langle \cdot, \cdot \rangle$ satisfies the following four properties. Let u, v, and w be vectors and α be a scalar, then:

1. $\langle u + v, w \rangle = \langle u, w \rangle + \langle v, w \rangle$.
2. $\langle \alpha v, w \rangle = \alpha \langle v, w \rangle$.
3. $\langle v, w \rangle = \langle w, v \rangle$.
4. $\langle v, v \rangle \geq 0$ and equal if and only if $v = 0$.

INTRODUCTION

In linear algebra, an inner product space is a vector space with an additional structure called an inner product. This additional structure associates each pair of vectors in the space with a scalar quantity known as the inner product of the vectors. Inner products allow the rigorous introduction of intuitive geometrical notions such as the length of a vector or the angle between two vectors. They also provide the means of defining orthogonality between vectors (zero inner product). Inner product spaces generalize Euclidean spaces (in which the inner product is the dot product, also known as the scalar product) to vector spaces of any (possibly infinite) dimension, and are studied in functional analysis.

An inner product naturally induces an associated norm, thus an inner product space is also a normed vector space. A complete space with an inner product is called a Hilbert space. An (incomplete) space with an inner product is called a pre-Hilbert space, since its completion with respect to the norm induced by the inner product is a Hilbert space. Inner product spaces over the field of complex numbers are sometimes referred to as unitary spaces.

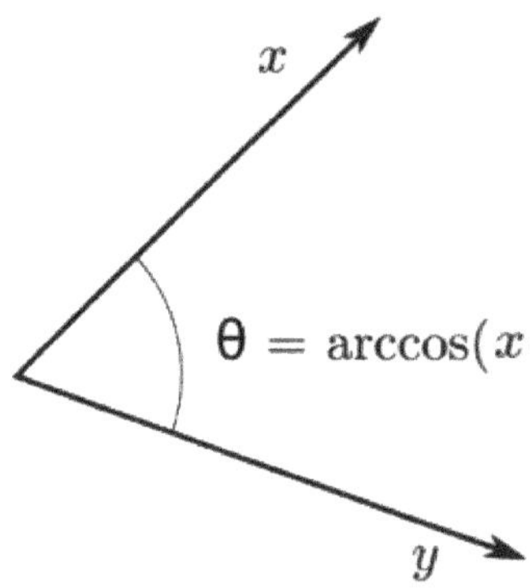

Geometric interpretation of the angle between two vectors defined using an inner product

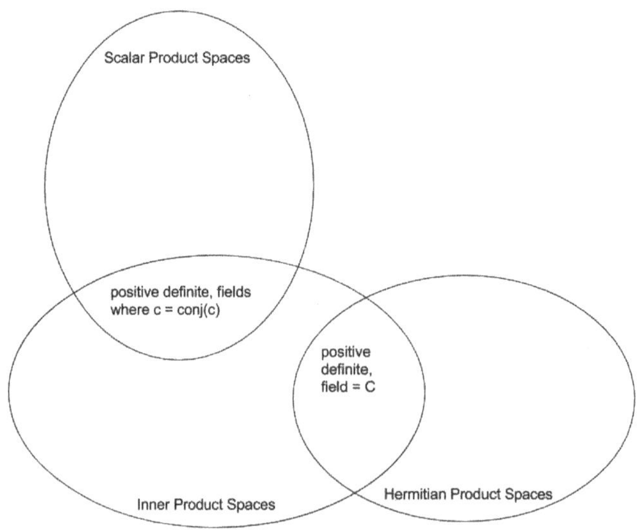

Scalar Product Spaces, over any field, have "scalar products" that are symmetrical and linear in the first(by convention) argument. Hermitian Product Spaces are restricted to the

field of complex numbers and have "hermitian products" that are conjugate-symmetrical and linear in the first argument(by convention). Inner Product Spaces may be defined over any field, having "inner products" that are linear in the first(by convention) argument, conjugate-symmetrical, and positive-definite. Unlike Inner Products, Scalar Products and Hermitian Products need not be positive-definite.

Definition

In this section, the field of scalars denoted F is either the field of real numbers **R** or the field of complex numbers **C**.

Formally, an inner product space is a vector space V over the field F together with an *inner product*, i.e., with a map

$$\langle\cdot,\cdot\rangle : V \times V \to F$$

that satisfies the following three axioms for all vectors $x, y, z \in V$ and all scalars $a \in F$:

- Conjugate symmetry:

$$\langle x, y\rangle = \overline{\langle y, x\rangle}$$

- Linearity in the first argument:

$$\langle ax, y\rangle = a\langle x, y\rangle$$
$$\langle x + y, z\rangle = \langle x, z\rangle + \langle y, z\rangle$$

- Positive-definiteness:

$$\langle x, x\rangle \geq 0$$
$$\langle x, x\rangle = 0 \Leftrightarrow x = \mathbf{0}.$$

Alternative Definitions, Notations and Remarks

Some authors, especially in physics and matrix algebra, prefer to define the inner product and the sesquilinear form with linearity in the second argument rather than the first. Then the first argument becomes conjugate linear, rather than the second. In those disciplines we would write the product $\langle x, y\rangle$ as $\langle y|x\rangle$ (the bra–ket notation of quantum mechanics), respectively $y^\dagger x$ (dot product as a case of the convention of forming the matrix product AB as the dot products of rows of $A$with columns of B). Here the kets and columns are identified with the vectors of V and the bras and rows with the dual vectors or linear functionals of the dual space V^*, with conjugacy associated with duality. This reverse order is now occasionally followed in the more abstract literature, taking $\langle x, y\rangle$ to be conjugate linear in x rather than y. A few instead find a middle ground by recognizing both $\langle\cdot,\cdot\rangle$ and $\langle\cdot|\cdot\rangle$ as distinct notations differing only in which argument is conjugate linear.

There are various technical reasons why it is necessary to restrict the basefield to **R** and **C** in the definition. Briefly, the basefield has to contain an ordered subfield in order for nonnegativity to make sense, and therefore has to have characteristic equal to 0 (since any ordered field has to have such characteristic). This immediately excludes finite fields. The basefield has to have additional structure, such as a distinguished automorphism. More generally any quadratically closed subfield of **R** or **C** will suffice for this purpose, e.g., the algebraic numbers, but when it is a proper subfield (i.e., neither **R** nor **C**) even finite-dimensional inner product spaces will fail to be metrically complete. In contrast all finite-dimensional inner product spaces over **R** or **C**, such as those used in quantum computation, are automatically metrically complete and hence Hilbert spaces.

In some cases we need to consider non-negative *semi-definite* sesquilinear forms. This means that $\langle x, x\rangle$ is only required to be non-negative. We show how to treat these below.

Elementary Properties

When $F = \mathbf{R}$, conjugate symmetry reduces to symmetry. That is, $\langle x, y\rangle = \langle y, x\rangle$ for $F = \mathbf{R}$; while for $F = \mathbf{C}$, $\langle x, y\rangle$ is equal to the complex conjugate.

Notice that conjugate symmetry implies that $\langle x, x\rangle$ is real for all x, since we have:

$$\langle x, x\rangle = \overline{\langle x, x\rangle}.$$

Moreover, sesquilinearity (see below) implies that

$$\langle -x, x\rangle = -1\langle x, x\rangle = \overline{-1}\langle x, x\rangle = \langle x, -x\rangle.$$

Conjugate symmetry and linearity in the first variable gives

$$\langle x, ay\rangle = \overline{\langle ay, x\rangle} = \bar{a}\overline{\langle y, x\rangle} = \bar{a}\langle x, y\rangle$$

$$\langle x, y+z\rangle = \overline{\langle y+z, x\rangle} = \overline{\langle y, x\rangle} + \overline{\langle z, x\rangle} = \langle x, y\rangle + \langle x, z\rangle,$$

so an inner product is a sesquilinear form. Conjugate symmetry is also called Hermitian symmetry, and a conjugate symmetric sesquilinear form is called a *Hermitian form*. While the above axioms are more mathematically economical, a compact verbal definition of an inner product is a *positive-definite Hermitian form*.

In the case of $F = \mathbf{R}$, conjugate-symmetry reduces to symmetry, and sesquilinear reduces to bilinear. So, an inner product on a real vector space is a *positive-definite symmetric bilinear form*.

From the linearity property it is derived that $x = 0$ implies $\langle x, x\rangle = 0$, while from the positive-definiteness axiom we obtain the converse, $\langle x, x\rangle = 0$ implies $x = 0$. Combining these two, we have the property that $\langle x, x\rangle = 0$ if and only if $x = 0$.

Combining the linearity of the inner product in its first argument and the conjugate symmetry gives the following important generalization of the familiar square expansion:

$$\langle x+y, x+y\rangle = \langle x, x\rangle + \langle x, y\rangle + \langle y, x\rangle + \langle y, y\rangle.$$

Assuming the underlying field to be $\mathbf{R}$, the inner product becomes symmetric, and we obtain

$$\langle x \pm y, x \pm y\rangle = \langle x, x\rangle \pm 2\langle x, y\rangle + \langle y, y\rangle,$$

The property of an inner product space V that

$$\langle x+y, z\rangle = \langle x, z\rangle + \langle y, z\rangle,$$

$$\langle x, y+z\rangle = \langle x, y\rangle + \langle x, z\rangle$$

is also known as *additivity*.

EXAMPLES

- A simple example is the real numbers with the standard multiplication as the inner product

$$\langle x, y\rangle := xy.$$

More generally, the real n-space $\mathbf{R}^n$ with the dot product is an inner product space, an example of a Euclidean n-space.

$$\left\langle \begin{bmatrix} x_1 \\ \vdots \\ x_n \end{bmatrix}, \begin{bmatrix} y_1 \\ \vdots \\ y_n \end{bmatrix} \right\rangle := x^{\mathrm{T}} y = \sum_{i=1}^{n} x_i y_i = x_1 y_1 + \cdots + x_n y_n,$$

where x^{T} is the transpose of x.

- The general form of an inner product on $\mathbf{C}^n$ is known as the Hermitian form and is given by

$$\langle \mathbf{x}, \mathbf{y}\rangle := \mathbf{y}^{\dagger}\mathbf{M}\mathbf{x} = \overline{\mathbf{x}^{\dagger}\mathbf{M}\mathbf{y}},$$

where $\mathbf{M}$ is any Hermitian positive-definite matrix and $\mathbf{y}^{\dagger}$ is the conjugate transpose of $\mathbf{y}$. For the real case this corresponds to the dot product of the results of directionally different scaling of the two vectors, with positive scale factors and orthogonal directions of scaling. Up to an orthogonal transformation it is a weighted-sum version of the dot product, with positive weights.

- The article on Hilbert space has several examples of inner product spaces wherein the metric induced by the inner product yields a complete metric space. An example of an inner product which induces an incomplete metric occurs with the space $C([a, b])$ of continuous complex valued functions on the interval $[a, b]$. The inner product is

$$\langle f, g\rangle := \int_a^b f(t)\overline{g(t)}\, dt.$$

This space is not complete; consider for example, for the interval $[-1,1]$ the sequence of continuous "step" functions, $\{f_k\}_k$, defined by:

$$f_k(t) = \begin{cases} 0 & t \in [-1, 0] \\ 1 & t \in \left[\frac{1}{k}, 1\right] \\ kt & t \in \left(0, \frac{1}{k}\right) \end{cases}$$

This sequence is a Cauchy sequence for the norm induced by the preceding inner product, which does not converge to a *continuous* function.

- For real random variables X and Y, the expected value of their product

$$\langle X, Y\rangle := \mathrm{E}(XY)$$

is an inner product. In this case, $\langle X, X\rangle = 0$ if and only if $\Pr(X = 0) = 1$ (i.e., $X = 0$ almost surely). This definition of expectation as inner product can be extended to random vectors as well.

- For real matrices of the same size, $\langle A, B\rangle := \mathrm{tr}(AB^{\mathrm{T}})$ with transpose as conjugation

$(\langle A, B\rangle = \langle B^{\mathrm{T}}, A^{\mathrm{T}}\rangle)$

is an inner product.

NORMS ON INNER PRODUCT SPACES

A linear space with a norm such as:

$$\|x\|_p = \left(\sum_{i=1}^{\infty} |\xi_i|^p\right)^{\frac{1}{p}} \qquad x = \{\xi_i\} \in \ell^p, p \neq 2,$$

is a normed space but not an inner product space, because this norm does not satisfy the parallelogram equality required of a norm to have an inner product associated with it.

However, inner product spaces have a naturally defined norm based upon the inner product of the space itself that does satisfy the parallelogram equality:

$$\|x\| = \sqrt{\langle x, x\rangle}.$$

This is well defined by the nonnegativity axiom of the definition of inner product space. The norm is thought of as the length of the vector x. Directly from the axioms, we can prove the following:

- Cauchy–Schwarz inequality: for x, y elements of V

$$|\langle x, y\rangle| \leq \|x\| \cdot \|y\|$$

with equality if and only if x and y are linearly dependent. This is one of the most important inequalities in mathematics. It is also known in the Russian mathematical literature as the *Cauchy–Bunyakovsky–Schwarz inequality*.

- Orthogonality: The geometric interpretation of the inner product in terms of angle and length, motivates much of the geometric terminology we use in regard to these spaces. Indeed, an immediate consequence of the Cauchy–Schwarz inequality is that it justifies defining the angle between two non-zero vectors x and y in the case $F = \mathbf{R}$ by the identity

$$\mathrm{angle}(x, y) = \arccos \frac{\langle x, y\rangle}{\|x\| \cdot \|y\|}.$$

We assume the value of the angle is chosen to be in the interval $[0, \pi]$. This is in analogy to the situation in two-dimensional Euclidean space.

In the case $F = \mathbf{C}$, the angle in the interval $[0, \pi/2]$ is typically defined by

$$\mathrm{angle}(x, y) = \arccos \frac{|\langle x, y\rangle|}{\|x\| \cdot \|y\|}.$$

Correspondingly, we will say that non-zero vectors x and y of V are orthogonal if and only if their inner product is zero.

- Homogeneity: for x an element of V and r a scalar

$$\|r \cdot x\| = |r| \cdot \|x\|.$$

The homogeneity property is completely trivial to prove.

- Triangle inequality: for x, y elements of V

$$\|x + y\| \leq \|x\| + \|y\|.$$

The last two properties show the function defined is indeed a norm.

Because of the triangle inequality and because of axiom 2, we see that $\|\cdot\|$ is a norm which turns V into a normed vector space and hence also into a metric space. The most important inner product spaces are the ones which are complete with respect to this metric; they are called Hilbert spaces. Every inner product V space is a dense subspace of some Hilbert space. This Hilbert space is essentially uniquely determined by V and is constructed by completing V.

- Pythagorean theorem: Whenever x, y are in V and $\langle x, y\rangle = 0$, then

$$\|x\|^2 + \|y\|^2 = \|x + y\|^2.$$

The proof of the identity requires only expressing the definition of norm in terms of the inner product and multiplying out, using the property of additivity of each component.

The name *Pythagorean theorem* arises from the geometric interpretation of this result as an analogue of the theorem in synthetic geometry. Note that the proof of the Pythagorean theorem in synthetic geometry is considerably more elaborate because of the paucity of underlying structure. In this sense, the synthetic Pythagorean theorem, if correctly demonstrated, is deeper than the version given above.

An induction on the Pythagorean theorem yields:

- If $x_1, \ldots, x_n$ are orthogonal vectors, that is, $\langle x_j, x_k\rangle = 0$ for distinct indices j, k, then

$$\sum_{i=1}^{n} \|x_i\|^2 = \left\|\sum_{i=1}^{n} x_i\right\|^2.$$

In view of the Cauchy-Schwarz inequality, we also note that $\langle\cdot,\cdot\rangle$ is continuous from $V \times V$ to F. This allows us to extend Pythagoras' theorem to infinitely many summands:

- Parseval's identity: Suppose V is a *complete* inner product space. If $\{x_k\}$ are mutually orthogonal vectors in V then

$$\sum_{i=1}^{\infty} \|x_i\|^2 = \left\|\sum_{i=1}^{\infty} x_i\right\|^2,$$

provided the infinite series on the left is convergent. Completeness of the space is needed to ensure that the sequence of partial sums

$$S_k = \sum_{i=1}^{k} x_i,$$

which is easily shown to be a Cauchy sequence, is convergent.

- Parallelogram law: for x, y elements of V,

$$\|x+y\|^2 + \|x-y\|^2 = 2\|x\|^2 + 2\|y\|^2.$$

The Parallelogram law is, in fact, a necessary and sufficient condition for the existence of a scalar product corresponding to a given norm. If it holds, the scalar product is defined by the polarization identity:

$$\|x+y\|^2 = \|x\|^2 + \|y\|^2 + 2\Re\langle x, y\rangle.$$

which is a form of the law of cosines.

ORTHONORMAL SEQUENCES

Let V be a finite dimensional inner product space of dimension n. Recall that every basis of V consists of exactly n linearly independent vectors. Using the Gram–Schmidt process we may start with an arbitrary basis and transform it into an orthonormal basis. That is, into a basis in which all the elements are orthogonal and have unit norm. In symbols, a basis $\{e_1, \ldots, e_n\}$ is orthonormal if $\langle e_i, e_j\rangle = 0$ if $i \neq j$ and $\langle e_i, e_i\rangle = \|e_i\| = 1$ for each i.

This definition of orthonormal basis generalizes to the case of infinite-dimensional inner product spaces in the following way. Let V be any inner product space. Then a collection

$$E = \{e_\alpha\}_{\alpha \in A}$$

is a *basis* for V if the subspace of V generated by finite linear combinations of elements of E is dense in V (in the norm induced by the inner product). We say that E is an *orthonormal basis* for V if it is a basis and

$$\langle e_\alpha, e_\beta\rangle = 0$$

if $\alpha \neq \beta$ and $\langle e_\alpha, e_\alpha\rangle = \|e_\alpha\| = 1$ for all $\alpha, \beta \in A$.

Using an infinite-dimensional analog of the Gram-Schmidt process one may show:

Theorem. Any separable inner product space V has an orthonormal basis.

Using the Hausdorff maximal principle and the fact that in a complete inner product space orthogonal projection onto linear subspaces is well-defined, one may also show that

Theorem. Any complete inner product space V has an orthonormal basis.

The two previous theorems raise the question of whether all inner product spaces have an orthonormal basis. The answer, it turns out is negative. This is a non-trivial result, and is proved below. The following proof is taken from Halmos's A Hilbert Space Problem Book (see the references).

Parseval's identity leads immediately to the following theorem:

Theorem. Let V be a separable inner product space and $\{e_k\}_k$ an orthonormal basis of V. Then the map

$$x \mapsto \{\langle e_k, x\rangle\}_{k\in\mathbb{N}}$$

is an isometric linear map $V \rightarrow l^2$ with a dense image.

This theorem can be regarded as an abstract form of Fourier series, in which an arbitrary orthonormal basis plays the role of the sequence of trigonometric polynomials. Note that the underlying index set can be taken to be any countable set (and in fact any set whatsoever, provided l^2 is defined appropriately, as is explained in the article Hilbert space). In particular, we obtain the following result in the theory of Fourier series:

Theorem. Let V be the inner product space $C[-\pi, \pi]$. Then the sequence (indexed on set of all integers) of continuous functions

$$e_k(t) = \frac{e^{ikt}}{\sqrt{2\pi}}$$

is an orthonormal basis of the space $C[-\pi, \pi]$ with the L^2 inner product. The mapping

$$f \mapsto \frac{1}{\sqrt{2\pi}}\left\{\int_{-\pi}^{\pi} f(t)e^{-ikt}\, dt\right\}_{k\in\mathbb{Z}}$$

is an isometric linear map with dense image.

Orthogonality of the sequence $\{e_k\}_k$ follows immediately from the fact that if $k \neq j$, then

$$\int_{-\pi}^{\pi} e^{-i(j-k)t}\, dt = 0.$$

Normality of the sequence is by design, that is, the coefficients are so chosen so that the norm comes out to 1. Finally the fact that the sequence has a dense algebraic span, in the *inner product norm*, follows from the fact that the sequence has a dense algebraic span, this time in the space of continuous periodic functions on $[-\pi, \pi]$ with the uniform norm. This is the content of the Weierstrass theorem on the uniform density of trigonometric polynomials.

Operators on Inner Product Spaces

Several types of linear maps A from an inner product space V to an inner product space W are of relevance:

- Continuous linear maps, i.e., A is linear and continuous with respect to the metric defined above, or equivalently, A is linear and the set of non-negative reals $\{||Ax||\}$, where x ranges over the closed unit ball of V, is bounded.
- Symmetric linear operators, i.e., A is linear and $\langle Ax, y\rangle = \langle x, Ay\rangle$ for all x, y in V.
- Isometries, i.e., A is linear and $\langle Ax, Ay\rangle = \langle x, y\rangle$ for all x, y in V, or equivalently, A is linear and $||Ax|| = ||x||$ for all x in V. All isometries are injective. Isometries are morphisms between inner product spaces, and morphisms of real inner product spaces are orthogonal transformations (compare with orthogonal matrix).
- Isometrical isomorphisms, i.e., A is an isometry which is surjective (and hence bijective). Isometrical isomorphisms are also known as unitary operators (compare with unitary matrix).

From the point of view of inner product space theory, there is no need to distinguish between two spaces which are isometrically isomorphic. The spectral theorem provides a canonical form for symmetric, unitary and more generally normal operators on finite dimensional inner product spaces. A generalization of the spectral theorem holds for continuous normal operators in Hilbert spaces.

GENERALIZATIONS

Any of the axioms of an inner product may be weakened, yielding generalized notions. The generalizations that are closest to inner products occur where bilinearity and conjugate symmetry are retained, but positive-definiteness is weakened.

Degenerate inner products

If V is a vector space and $\langle\cdot,\cdot\rangle$ a semi-definite sesquilinear form, then the function:

$$\|x\| = \sqrt{\langle x, x\rangle}$$

makes sense and satisfies all the properties of norm except that $\|x\| = 0$ does not imply $x = 0$ (such a functional is then called a semi-norm). We can produce an inner product space by considering the quotient $W = V/\{ x : \|x\| = 0\}$. The sesquilinear form $\langle\cdot,\cdot\cdot\cdot\rangle$ factors through W.

This construction is used in numerous contexts. The Gelfand–Naimark–Segal construction is a particularly important example of the use of this technique. Another example is the representation of semi-definite kernels on arbitrary sets.

Nondegenerate conjugate symmetric forms

Alternatively, one may require that the pairing be a nondegenerate form, meaning that for all non-zero x there exists some $y$such that $\langle x, y\rangle \neq 0$, though y need not equal x; in other words, the induced map to the dual space $V \rightarrow V^*$ is injective. This generalization is important in differential geometry: a manifold whose tangent spaces have an inner product is a Riemannian manifold, while if this is related to nondegenerate conjugate symmetric form the manifold is a pseudo-Riemannian manifold. By Sylvester's law of inertia, just as every inner product is similar to the dot product with positive weights on a set of vectors, every nondegenerate conjugate symmetric form is similar to the dot product with *nonzero* weights on a set of vectors, and the number of positive and negative weights are called respectively the positive index and negative index. Product of vectors in Minkowski space is an example of indefinite inner product, although, technically speaking, it is not an inner product according to the standard definition above. Minkowski space has four dimensions and indices 3 and 1 (assignment of "+" and "−" to them differs depending on conventions).

Purely algebraic statements (ones that do not use positivity) usually only rely on the nondegeneracy (the injective homomorphism $V \rightarrow V^*$) and thus hold more generally.

Related products

The term "inner product" is opposed to outer product, which is a slightly more general opposite. Simply, in coordinates, the inner product is the product of a $1\times n$ *co*vector with an $n\times 1$ vector, yielding a 1×1 matrix (a scalar), while the outer product is the product of an $m\times 1$

vector with a $1 \times n$ covector, yielding an $m \times n$ matrix. Note that the outer product is defined for different dimensions, while the inner product requires the same dimension. If the dimensions are the same, then the inner product is the *trace* of the outer product (trace only being properly defined for square matrices).

On an inner product space, or more generally a vector space with a nondegenerate form (so an isomorphism $V \rightarrow V^*$) vectors can be sent to covectors (in coordinates, via transpose), so one can take the inner product and outer product of two vectors, not simply of a vector and a covector.

In a quip: "inner is horizontal times vertical and shrinks down, outer is vertical times horizontal and expands out".

More abstractly, the outer product is the bilinear map $W \times V^* \rightarrow \mathbf{Hom}(V, W)$ sending a vector and a covector to a rank 1 linear transformation (simple tensor of type (1,1)), while the inner product is the bilinear evaluation map $V^* \times V \rightarrow F$ given by evaluating a covector on a vector; the order of the domain vector spaces here reflects the covector/vector distinction.

The inner product and outer product should not be confused with the interior product and exterior product, which are instead operations on vector fields and differential forms, or more generally on the exterior algebra.

As a further complication, in geometric algebra the inner product and the *exterior* (Grassmann) product are combined in the geometric product (the Clifford product in a Clifford algebra) – the inner product sends two vectors (1-vectors) to a scalar (a 0-vector), while the exterior product sends two vectors to a bivector (2-vector) – and in this context the exterior product is usually called the "*outer* (alternatively, wedge) product". The inner product is more correctly called a *scalar* product in this context, as the nondegenerate quadratic form in question need not be positive definite (need not be an inner product).

REFERENCES

1. P. K. Jain, Khalil Ahmad (1995). "5.1 Definitions and basic properties of inner product spaces and Hilbert spaces". Functional analysis (2nd ed.). New Age International. p. 203. ISBN 81-224-0801-X.
2. Eduard Prugovecki (1981). "Definition 2.1". Quantum mechanics in Hilbert space (2nd ed.). Academic Press. pp. 18 ff. ISBN 0-12-566060-X.
3. Emch, Gerard G. (1972). Algebraic methods in statistical mechanics and quantum field theory. New York: Wiley-Interscience. ISBN 978-0-471-23900-0.
4. Finkbeiner, Daniel T. (2013), Introduction to Matrices and Linear Transformations, Dover Books on Mathematics (3rd ed.), Courier Dover Publications, p. 242, ISBN 9780486279664.
5. P. K. Jain, Khalil Ahmad (1995). "Example 5". Cited work. p. 209. ISBN 81-224-0801-X.
6. Karen Saxe (2002). Beginning functional analysis. Springer. p. 7. ISBN 0-387-95224-1.

CHAPTER 9

Category Theory

Chapter Contents

A category consists of three things: a collection of objects, for each pair of objects a collection of morphisms (sometimes call "arrows") from one to another, and a binary operation defined on compatible pairs of morphisms called composition. The category must satisfy an identity axiom and an associative axiom which is analogous to the monoid axioms.

INTRODUCTION

Category theory formalizes mathematical structure and its concepts in terms of a collection of objects and of arrows (also called morphisms). A category has two basic properties: the ability to compose the arrows associatively and the existence of an identity arrow for each object. The language of category theory has been used to formalize concepts of other high-level abstractions such as sets, rings, and groups.

Several terms used in category theory, including the term "morphism", are used differently from their uses in the rest of mathematics. In category theory, morphisms obey conditions specific to category theory itself. Samuel Eilenberg and Saunders Mac Lane introduced the concepts of categories, functors, and natural transformations in 1942–45 in their study of algebraic topology, with the goal of understanding the processes that preserve mathematical structure.

Category theory has practical applications in programming language theory, in particular for the study of monads in functional programming

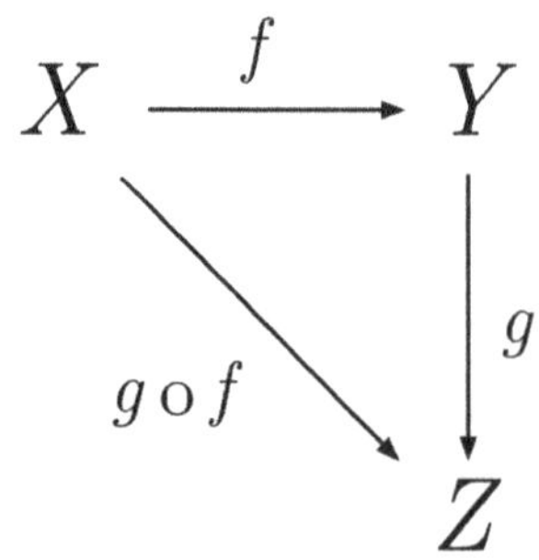

Figure 1: Schematic representation of a category with objects X, Y, Z and morphisms f, g, g ∘ f. (The category's three identity morphisms 1_X, 1_Y and 1_Z, if explicitly represented, would appear as three arrows, next to the letters X, Y, and Z, respectively, each having as its "shaft" a circular arc measuring almost 360 degrees.)

BASIC CONCEPTS

Categories represent abstraction of other mathematical concepts. Many areas of mathematics can be formalised by category theory as categories. Hence category theory uses abstraction to make it possible to state and prove many intricate and subtle mathematical results in these fields in a much simpler way.

A basic example of a category is the category of sets, where the objects are sets and the arrows are functions from one set to another. However, the objects of a category need not be sets, and the arrows need not be functions. Any way of formalising a mathematical concept such that it meets the basic conditions on the behaviour of objects and arrows is a valid category—and all the results of category theory apply to it.

The "arrows" of category theory are often said to represent a process connecting two objects, or in many cases a "structure-preserving" transformation connecting two objects. There are, however, many applications where much more abstract concepts are represented by objects and morphisms. The most important property of the arrows is that they can be "composed", in other words, arranged in a sequence to form a new arrow.

APPLICATIONS OF CATEGORIES

Categories now appear in many branches of mathematics, some areas of theoretical computer science where they can correspond to types or to database schemas, and mathematical physics where they can be used to describe vector spaces. Linear algebra can also be expressed in terms of categories of matrices.

UTILITY

Categories, objects, and morphisms

The study of categories is an attempt to axiomatically capture what is commonly found in various classes of related mathematical structures by relating them to the structure-preserving functions between them. A systematic study of category theory then allows us to prove general results about any of these types of mathematical structures from the axioms of a category.

Consider the following example. The class Grp of groups consists of all objects having a "group structure". One can proceed to prove theorems about groups by making logical deductions from the set of axioms. For example, it is immediately proven from the axioms that the identity element of a group is unique.

Instead of focusing merely on the individual objects (e.g., groups) possessing a given structure, category theory emphasizes the morphisms – the structure-preserving mappings – between these objects; by studying these morphisms, we are able to learn more about the structure of the objects. In the case of groups, the morphisms are the group homomorphisms. A group homomorphism between two groups "preserves the group structure" in a precise sense – it is a "process" taking one group to another, in a way that carries along information about the structure of the first group into the second group. The study of group homomorphisms then provides a tool for studying general properties of groups and consequences of the group axioms.

A similar type of investigation occurs in many mathematical theories, such as the study of continuous maps (morphisms) between topological spaces in topology (the associated category is called Top), and the study of smooth functions (morphisms) in manifold theory.

Not all categories arise as "structure preserving (set) functions", however; the standard example is the category of homotopies between pointed topological spaces.

If one axiomatizes relations instead of functions, one obtains the theory of allegories.

Functors

A category is itself a type of mathematical structure, so we can look for "processes" which preserve this structure in some sense; such a process is called a functor.

Diagram chasing is a visual method of arguing with abstract "arrows" joined in diagrams. Functors are represented by arrows between categories, subject to specific defining commutativity conditions. Functors can define (construct) categorical diagrams and sequences (viz. Mitchell, 1965). A functor associates to every object of one category an object of another category, and to every morphism in the first category a morphism in the second.

In fact, what we have done is define a category of categories and functors – the objects are categories, and the morphisms (between categories) are functors.

By studying categories and functors, we are not just studying a class of mathematical structures and the morphisms between them; we are studying the relationships between various classes of mathematical structures. This is a fundamental idea, which first surfaced in algebraic topology. Difficult topological questions can be translated into algebraic questions which are often easier to solve. Basic constructions, such as the fundamental group or the fundamental groupoid of a topological space, can be expressed as functors to the category of groupoids in this way, and the concept is pervasive in algebra and its applications.

Natural Transformations

Abstracting yet again, some diagrammatic and/or sequential constructions are often "naturally related" – a vague notion, at first sight. This leads to the clarifying concept of natural transformation, a way to "map" one functor to another. Many important constructions in mathematics can be studied in this context. "Naturality" is a principle, like general covariance in physics, that cuts deeper than is initially apparent. An arrow between two functors is a natural transformation when it is subject to certain naturality or commutativity conditions.

Functors and natural transformations ('naturality') are the key concepts in category theory

CATEGORIES, OBJECTS, AND MORPHISMS

Categories

A category C consists of the following three mathematical entities:

- A class ob(C), whose elements are called objects;
- A class hom(C), whose elements are called morphisms or maps or arrows. Each morphism f has a source object a and target object b.

  The expression $f : a \to b$, would be verbally stated as "f is a morphism from a to b".

  The expression hom(a, b) — alternatively expressed as homC(a, b), mor(a, b), or C(a, b) — denotes the hom-class of all morphisms from a to b.
- A binary operation $\circ$, called composition of morphisms, such that for any three objects a, b, and c, we have $\text{hom}(b, c) \times \text{hom}(a, b) \to \text{hom}(a, c)$. The composition of $f : a \to b$ and $g : b \to c$ is written as $g \circ f$ or gf, governed by two axioms:
  - Associativity: If $f : a \to b$, $g : b \to c$ and $h : c \to d$ then $h \circ (g \circ f) = (h \circ g) \circ f$, and
  - Identity: For every object x, there exists a morphism $1x : x \to x$ called the identity morphism for x, such that for every morphism $f : a \to b$, we have $1b \circ f = f = f \circ 1a$.

From the axioms, it can be proved that there is exactly one identity morphism for every object. Some authors deviate from the definition just given by identifying each object with its identity morphism.

Morphisms

Relations among morphisms (such as fg = h) are often depicted using commutative diagrams, with "points" (corners) representing objects and "arrows" representing morphisms.

Morphisms can have any of the following properties. A morphism $f : a \to b$ is a:

- monomorphism (or monic) if $f \circ g_1 = f \circ g_2$ implies $g_1 = g_2$ for all morphisms $g_1, g_2 : x \to a$.
- epimorphism (or epic) if $g_1 \circ f = g_2 \circ f$ implies $g_1 = g_2$ for all morphisms g_1, $g_2 : b \to x$.
- bimorphism if f is both epic and monic.
- isomorphism if there exists a morphism $g : b \to a$ such that $f \circ g = 1_b$ and $g \circ f = 1_a$.
- endomorphism if a = b. end(a) denotes the class of endomorphisms of a.
- automorphism if f is both an endomorphism and an isomorphism. aut(a) denotes the class of automorphisms of a.
- retraction if a right inverse of f exists, i.e. if there exists a morphism $g : b \to a$ with $f \circ g = 1_b$.
- section if a left inverse of f exists, i.e. if there exists a morphism $g : b \to a$ with $g \circ f = 1_a$.

Every retraction is an epimorphism, and every section is a monomorphism. Furthermore, the following three statements are equivalent:

- f is a monomorphism and a retraction;
- f is an epimorphism and a section;
- f is an isomorphism.

FUNCTORS

Functors are structure-preserving maps between categories. They can be thought of as morphisms in the category of all (small) categories.

A (covariant) functor F from a category C to a category D, written $F : C \to D$, consists of:

- for each object x in C, an object F(x) in D; and
- for each morphism $f : x \to y$ in C, a morphism $F(f) : F(x) \to F(y)$,

such that the following two properties hold:

- For every object x in C, $F(1_x) = 1_{F(x)}$;

- For all morphisms f : x → y and g : y → z, F(g ∘ f) = F(g) ∘ F(f).

A contravariant functor F: C → D, is like a covariant functor, except that it "turns morphisms around" ("reverses all the arrows"). More specifically, every morphism f : x → y in C must be assigned to a morphism F(f) : F(y) → F(x) in D. In other words, a contravariant functor acts as a covariant functor from the opposite category Cop to D.

NATURAL TRANSFORMATIONS

A natural transformation is a relation between two functors. Functors often describe "natural constructions" and natural transformations then describe "natural homomorphisms" between two such constructions. Sometimes two quite different constructions yield "the same" result; this is expressed by a natural isomorphism between the two functors.

If F and G are (covariant) functors between the categories C and D, then a natural transformation η from F to G associates to every object X in C a morphism ηX : F(X) → G(X) in D such that for every morphism f : X → Y in C, we have ηY ∘ F(f) = G(f) ∘ η_X; this means that the following diagram is commutative:

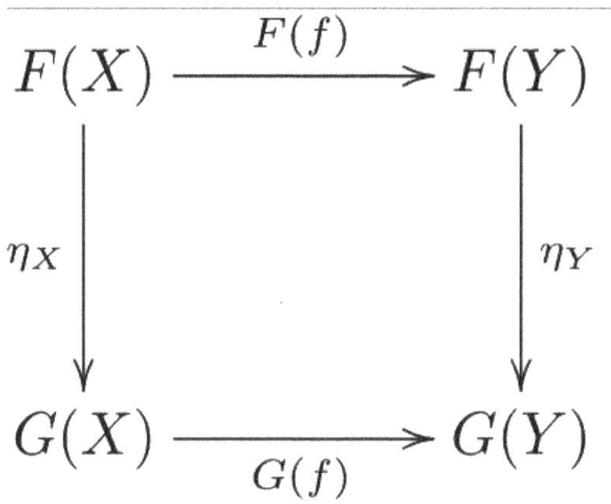

The two functors F and G are called naturally isomorphic if there exists a natural transformation from F to G such that η_X is an isomorphism for every object X in C.

OTHER CONCEPTS

Universal constructions, limits, and colimits

Using the language of category theory, many areas of mathematical study can be categorized. Categories include sets, groups and topologies.

Each category is distinguished by properties that all its objects have in common, such as the empty set or the product of two topologies, yet in the definition of a category, objects are considered atomic, i.e., we do not know whether an object A is a set, a topology, or any other abstract concept. Hence, the challenge is to define special objects without referring to the internal structure of those objects. To define the empty set without referring to elements, or the product topology without referring to open sets, one can characterize these objects in terms of their relations to other objects, as given by the morphisms of the respective categories. Thus, the task is to find universal properties that uniquely determine the objects of interest.

Indeed, it turns out that numerous important constructions can be described in a purely categorical way. The central concept which is needed for this purpose is called categorical limit, and can be dualized to yield the notion of a colimit.

Equivalent categories

It is a natural question to ask: under which conditions can two categories be considered essentially the same, in the sense that theorems about one category can readily be transformed into theorems about the other category? The major tool one employs to describe such a situation is called equivalence of categories, which is given by appropriate functors between two categories. Categorical equivalence has found numerous applications in mathematics.

Further concepts and results

The definitions of categories and functors provide only the very basics of categorical algebra; additional important topics are listed below. Although there are strong interrelations between all of these topics, the given order can be considered as a guideline for further reading.

- The functor category D^C has as objects the functors from C to D and as morphisms the natural transformations of such functors. The Yoneda lemma is one of the most famous basic results of category theory; it describes representable functors in functor categories.
- Duality: Every statement, theorem, or definition in category theory has a dual which is essentially obtained by "reversing all the arrows". If one statement is true in a category C then its dual is true in the dual category Cop. This duality, which is transparent at the level of category theory, is often obscured in applications and can lead to surprising relationships.
- Adjoint functors: A functor can be left (or right) adjoint to another functor that maps in the opposite direction. Such a pair of adjoint functors typically arises from a construction defined by a universal property; this can be seen as a more abstract and powerful view on universal properties.

HIGHER-DIMENSIONAL CATEGORIES

Many of the above concepts, especially equivalence of categories, adjoint functor pairs, and functor categories, can be situated into the context of higher-dimensional categories. Briefly, if we consider a morphism between two objects as a "process taking us from one object to another", then higher-dimensional categories allow us to profitably generalize this by considering "higher-dimensional processes".

For example, a (strict) 2-category is a category together with "morphisms between morphisms", i.e., processes which allow us to transform one morphism into another. We can then "compose" these "bimorphisms" both horizontally and vertically, and we require a 2-dimensional "exchange law" to hold, relating the two composition laws. In this context, the standard example is Cat, the 2-category of all (small) categories, and in this example, bimorphisms of morphisms are simply natural transformations of morphisms in the usual sense. Another basic example is to consider a 2-category with a single object; these are essentially monoidal categories. Bicategories are a weaker notion of 2-dimensional categories in which the composition of morphisms is not strictly associative, but only associative "up to" an isomorphism.

This process can be extended for all natural numbers n, and these are called n-categories. There is even a notion of ω-category corresponding to the ordinal number ω.

Higher-dimensional categories are part of the broader mathematical field of higher-dimensional algebra, a concept introduced by Ronald Brown. For a conversational introduction to these ideas, see John Baez, 'A Tale of n-categories' (1996)

HISTORICAL NOTES

In 1942–45, Samuel Eilenberg and Saunders Mac Lane introduced categories, functors, and natural transformations as part of their work in topology, especially algebraic topology. Their work was an important part of the transition from intuitive and geometric homology to axiomatic homology theory. Eilenberg and Mac Lane later wrote that their goal was to understand natural transformations. That required defining functors, which required categories.

Stanislaw Ulam, and some writing on his behalf, have claimed that related ideas were current in the late 1930s in Poland. Eilenberg was Polish, and studied mathematics in Poland in the 1930s. Category theory is also, in some sense, a continuation of the work of Emmy Noether (one of Mac Lane's teachers) in formalizing abstract processes; Noether realized that understanding a type of mathematical structure requires understanding the processes that preserve that structure. To achieve this understanding, Eilenberg and Mac Lane proposed an axiomatic formalization of the relation between structures and the processes that preserve them.

The subsequent development of category theory was powered first by the computational needs of homological algebra, and later by the axiomatic needs of algebraic geometry, the field most resistant to being grounded in either axiomatic set theory or the Russell-Whitehead view of united foundations. General category theory, an extension of universal algebra having many new features allowing for semantic flexibility and higher-order logic, came later; it is now applied throughout mathematics.

Certain categories called topoi (singular topos) can even serve as an alternative to axiomatic set theory as a foundation of mathematics. A topos can also be considered as a specific type of category with two additional topos axioms. These foundational applications of category theory have been worked out in fair detail as a basis for, and justification of, constructive mathematics. Topos theory is a form of abstract sheaf theory, with geometric origins, and leads to ideas such as pointless topology.

Categorical logic is now a well-defined field based on type theory for intuitionistic logics, with applications in functional programming and domain theory, where a cartesian closed category is taken as a non-syntactic description of a lambda calculus. At the very least, category theoretic language clarifies what exactly these related areas have in common (in some abstract sense).

Category theory has been applied in other fields as well. For example, John Baez has shown a link between Feynman diagrams in Physics and monoidal categories. Another application of category theory, more specifically: topos theory, has been made in mathematical music theory, see for example the book The Topos of Music, Geometric Logic of Concepts, Theory, and Performance by Guerino Mazzola.

More recent efforts to introduce undergraduates to categories as a foundation for mathematics include those of William Lawvere and Rosebrugh (2003) and Lawvere and Stephen Schanuel (1997) and Mirroslav Yotov (2012).

REFERENCES

1. Adámek, Jiří; Herrlich, Horst; Strecker, George E. (2004). Abstract and Concrete Categories. Heldermann Verlag Berlin.
2. Barr, Michael; Wells, Charles (2012), Category Theory for Computing Science, Reprints in Theory and Applications of Categories, 22 (3rd ed.).
3. Barr, Michael; Wells, Charles (2005), Toposes, Triples and Theories, Reprints in Theory and Applications of Categories, 12 (revised ed.), MR 2178101.
4. Borceux, Francis (1994). Handbook of categorical algebra. Encyclopedia of Mathematics and its Applications 50-52. Cambridge University Press.
5. Freyd, Peter J. (1964). Abelian Categories. New York: Harper and Row.
6. Freyd, Peter J.; Scedrov, Andre (1990). Categories, allegories. North Holland Mathematical Library. 39. North Holland. ISBN 978-0-08-088701-2.
7. Goldblatt, Robert (2006) [1979]. Topoi: The Categorial Analysis of Logic. Studies in logic and the foundations of mathematics. 94 (Reprint, revised ed.). Dover Publications. ISBN 978-0-486-45026-1.
8. Herrlich, Horst; Strecker, George E. (2007), Category Theory (3rd ed.), Heldermann Verlag Berlin, ISBN 978-3-88538-001-6.
9. Kashiwara, Masaki; Schapira, Pierre (2006). Categories and Sheaves. Grundlehren der Mathematischen Wissenschaften. 332. Springer. ISBN 978-3-540-27949-5.
10. Lawvere, F. William; Rosebrugh, Robert (2003). Sets for Mathematics. Cambridge University Press. ISBN 978-0-521-01060-3.
11. Lawvere, F. W.; Schanuel, Stephen Hoel (2009) [1997]. Conceptual Mathematics: A First Introduction to Categories (2nd ed.). Cambridge University Press. ISBN 978-0-521-89485-2.
12. Leinster, Tom (2004). Higher Operads, Higher Categories. London Math. Society Lecture Note Series. 298. Cambridge University Press. ISBN 978-0-521-53215-0.

13. Leinster, Tom (2014). Basic Category Theory. Cambridge University Press.
14. Lurie, Jacob (2009). Higher Topos Theory. Annals of Mathematics Studies. 170. Princeton, NJ: Princeton University Press. arXiv:math.CT/0608040Freely accessible. ISBN 978-0-691-14049-0. MR 2522659.
15. Mac Lane, Saunders (1998). Categories for the Working Mathematician. Graduate Texts in Mathematics. 5 (2nd ed.). Springer-Verlag. ISBN 0-387-98403-8. MR 1712872.
16. Mac Lane, Saunders; Birkhoff, Garrett (1999) [1967]. Algebra (2nd ed.). Chelsea. ISBN 0-8218-1646-2.
17. Martini, A.; Ehrig, H.; Nunes, D. (1996). "Elements of basic category theory". Technical Report. Technical University Berlin. 96 (5).
18. May, Peter (1999). A Concise Course in Algebraic Topology. University of Chicago Press. ISBN 0-226-51183-9.
19. Guerino, Mazzola (2002). The Topos of Music, Geometric Logic of Concepts, Theory, and Performance. Birkhäuser. ISBN 3-7643-5731-2.
20. Pedicchio, Maria Cristina; Tholen, Walter, eds. (2004). Categorical foundations. Special topics in order, topology, algebra, and sheaf theory. Encyclopedia of Mathematics and Its Applications. 97. Cambridge: Cambridge University Press. ISBN 0-521-83414-7. Zbl 1034.18001.
21. Pierce, Benjamin C. (1991). Basic Category Theory for Computer Scientists. MIT Press. ISBN 978-0-262-66071-6.
22. Schalk, A.; Simmons, H. (2005). An introduction to Category Theory in four easy movements (PDF). Notes for a course offered as part of the MSc. in Mathematical Logic, Manchester University.
23. Simpson, Carlos. Homotopy theory of higher categories. arXiv:1001.4071Freely accessible., draft of a book.
24. Taylor, Paul (1999). Practical Foundations of Mathematics. Cambridge Studies in Advanced Mathematics. 59. Cambridge University Press. ISBN 978-0-521-63107-5.
25. Turi, Daniele (1996–2001). "Category Theory Lecture Notes" (PDF). Retrieved 11 December 2009. Based on Mac Lane 1998.

CHAPTER 10

List Of Abstract Algebra Topics

Chapter Contents

Abstract algebra is the set of advanced topics of algebra that deal with abstract algebraic structures rather than the usual number systems. The most important of these structures are groups, rings, and fields. Important branches of abstract algebra are commutative algebra, representation theory, and homological algebra.

INTRODUCTION

Abstract algebra is the subject area of mathematics that studies algebraic structures, such as groups, rings, fields, modules, vector spaces, and algebras. The phrase abstract algebra was coined at the turn of the 20th century to distinguish this area from what was normally referred to as algebra, the study of the rules for manipulating formulae and algebraic expressions involving unknowns and real or complex numbers, often now called elementary algebra. The distinction is rarely made in more recent writings.

BASIC LANGUAGE

Algebraic structures are defined primarily as sets with *operations*.

- Algebraic structure
    - Subobjects: subgroup, subring, subalgebra, submodule etc.

- Binary operation
    - Closure of an operation
    - Associative property
    - Distributive property
    - Commutative property
- Unary operator
    - Additive inverse, multiplicative inverse, inverse element
        - Identity element
        - Cancellation property
- Finitary operation
    - Arity

Structure preserving maps called *homomorphisms* are vital in the study of algebraic objects.

- Homomorphisms
    - Kernels and cokernels
    - Image and coimage
    - Epimorphisms and monomorphisms
        - Isomorphisms
            - Isomorphism theorems

There are several basic ways to combine algebraic objects of the same type to produce a third object of the same type. These constructions are used throughout algebra.

- Direct sum
    - Direct limit
- Direct product
    - Inverse limit
- Quotient objects: quotient group, quotient ring, quotient module etc.
- Tensor product

Advanced concepts:

- Category theory
    - Category of groups
        - Category of abelian groups
    - Category of rings
    - Category of modules
        - Morita equivalence, Morita duality
    - Category of vector spaces
- Homological algebra
    - Filtration (algebra)
    - Exact sequence
    - Functor
- Zorn's lemma

Semigroups and Monoids

- Semigroup
    - Subsemigroup
    - Free semigroup
    - Green's relations
    - Inversion semigroup (cf.)
    - Krohn–Rhodes theory
    - Semigroup algebra
    - Transformation semigroup
- Monoid
    - Aperiodic monoid
    - Free monoid
    - Monoid (category theory)

    - Monoid factorisation
    - Syntactic monoid

Group Theory

Structure

- Group (mathematics)
    - Lagrange's theorem (group theory)
    - Subgroup
        - Coset
        - Normal subgroup
        - Characteristic subgroup
        - Centralizer and normalizer subgroups
        - Derived group
        - Frattini subgroup
        - Fitting subgroup
    - Classification of finite simple groups
    - Sylow theorems
    - Local analysis

Constructions

- Free group
    - Presentation of a group
    - Word problem for groups
- Quotient group
    - Extension problem
- Direct sum, direct product
- Semidirect product

  - Wreath product

Types

- Simple group
- Finite group
- Abelian group
  - Torsion subgroup
  - Free abelian group
  - Finitely generated abelian group
  - Rank of an abelian group
- Cyclic group
  - Locally cyclic group
- Solvable group
  - Composition series
- Nilpotent group
- Divisible group
- Dedekind group, Hamiltonian group

Examples

- Examples of groups
  - Permutation group
  - Symmetric group
  - Alternating group
  - p-group
  - List of small groups
  - Klein four-group
  - Quaternion group

    - Dihedral group
    - Dicyclic group
    - Automorphism group
    - Point group

Applications

- Group action
    - Conjugacy class
    - Inner automorphism
    - Conjugate closure
    - Stabilizer subgroup
    - Orbit (group theory)
    - Orbit-stabilizer theorem
    - Cayley's theorem
    - Burnside's lemma
- Burnside's problem
- Loop group
- Fundamental group

Ring Theory

General

- Ring (mathematics)
- Commutative algebra, Commutative ring
- Ring theory, Noncommutative ring
- Algebra over a field
    - Non-associative algebra

- Relatives to rings: Semiring, Nearring, Rig (algebra)

Structure

- Subring, Subalgebra
    - Center (algebra)
- Ring ideal
    - Principal ideal
    - Ideal quotient
    - Maximal ideal, minimal ideal
    - Primitive ideal, prime ideal, semiprime ideal
    - Radical of an ideal
- Jacobson radical
- Socle of a ring
- unit (ring theory), Idempotent, Nilpotent, Zero divisor
- Characteristic (algebra)
- Ring homomorphism, Algebra homomorphism
    - Ring epimorphism
    - Ring monomorphism
    - Ring isomorphism
        - Skolem–Noether theorem
- Graded algebra
- Morita equivalence
    - Brauer group

Constructions

- Direct sum of rings, Product of rings
- Quotient ring

- Matrix ring
- Endomorphism ring
- Polynomial ring
- Formal power series
- Monoid ring, Group ring, Group algebra
- Localization of a ring
- Tensor algebra
    - Symmetric algebra, Exterior algebra, Clifford algebra
- Free algebra
- Completion (ring theory)

Types

- Field (mathematics), Division ring, division algebra
- Simple ring, Central simple algebra, Semisimple ring, Semisimple algebra
- Primitive ring, Semiprimitive ring
- Prime ring, Semiprime ring, Reduced ring
- Integral domain, Domain (ring theory)
    - Field of fractions, Integral closure
    - Euclidean domain, Principal ideal domain, Unique factorization domain, Dedekind domain, Prüfer domain
- Von Neumann regular ring
- Quasi-Frobenius ring
- Hereditary ring, Semihereditary ring
- Local ring, Semi-local ring
- Discrete valuation ring
- Regular local ring
- Cohen–Macaulay ring

- Gorenstein ring
- Artinian ring, Noetherian ring
- Perfect ring, semiperfect ring
- Baer ring, Rickart ring
- Lie ring, Lie algebra
    - Ideal (Lie algebra)
- Jordan algebra
- Differential algebra
- Banach algebra

Examples

- Rational number, Real number, Complex number, Quaternions, Octonions
    - Hurwitz quaternion
- Gaussian integer

Theorems and applications

- Algebraic geometry
    - Hilbert's Nullstellensatz
- Hilbert's basis theorem
- Hopkins–Levitzki theorem
- Krull's principal ideal theorem
- Levitzky's theorem
- Galois theory
    - Abel–Ruffini theorem
- Artin-Wedderburn theorem
- Jacobson density theorem

- Wedderburn's little theorem
- Lasker–Noether theorem

FIELD THEORY

Basic concepts

- Field (mathematics)
- Subfield (mathematics)
    - Multiplicative group
        - Primitive element (field theory)
- Field extension
    - Algebraic extension
        - Splitting field
        - Algebraically closed field
            - Algebraic element
            - Algebraic closure
    - Separable extension
        - Separable polynomial
    - Normal extension
    - Galois extension
    - Abelian extension
    - Transcendence degree
- Field norm
- Field trace
- Conjugate element (field theory)
- Tensor product of fields

Types

- Algebraic number field
- Global field
- Local field
- Finite field
- Symmetric function
- Formally real field
- Real closed field

Applications

- Galois theory
    - Galois group
    - Inverse Galois problem
- Kummer theory

Module Theory

General

- Module (mathematics)
- Bimodule
- Annihilator (ring theory)

Structure

- Submodule
    - Pure submodule
- Module homomorphism
    - Essential submodule
    - Superfluous submodule

    - Singular submodule
- Socle of a module
- Radical of a module

Constructions

- Free module
- Quotient module
- Direct sum, Direct product of modules
- Direct limit, Inverse limit
- Localization of a module
- Completion (ring theory)

Types

- Simple module, Semisimple module
- Indecomposable module
- Artinian module, Noetherian module
- Homological types:
    - Projective module
        - Projective cover
        - Swan's theorem
        - Quillen–Suslin theorem
    - Injective module
        - Injective hull
    - Flat module
        - Flat cover
- Coherent module
- Finitely-generated module

- Finitely-presented module
- Finitely related module
- Algebraically compact module
- Reflexive module

Concepts and theorems

- Composition series
    - Length of a module
- Structure theorem for finitely generated modules over a principal ideal domain
- Homological dimension
    - Projective dimension
    - Injective dimension
    - Flat dimension
    - Global dimension
    - Weak global dimension
    - Cohomological dimension
- Krull dimension
- Regular sequence (algebra), depth (algebra)
- Fitting lemma
- Schur's lemma
- Nakayama's lemma
- Krull–Schmidt theorem
- Steinitz exchange lemma
- Jordan–Hölder theorem
- Artin–Rees lemma
- Schanuel's lemma

- Morita equivalence
    - Progenerator

REPRESENTATION THEORY

- Algebra representation
- Group representation
- Lie algebra representation
- Maschke's theorem
- Schur's lemma
- Equivariant map
- Affine representation
- Projective representation
- Modular representation theory
- Quiver (mathematics)
- Representation theory of Hopf algebras

Non-Associative Systems

General

- Associative property, Associator
- Magma (algebra)
    - Loop (algebra), Quasigroup
- Nonassociative ring, Non-associative algebra
    - Universal enveloping algebra
    - Lie algebra (*see also list of Lie group topics* and list of representation theory topics)
    - Jordan algebra
    - Alternative algebra

    - Power associativity
    - Flexible algebra

Examples

- Cayley–Dickson construction
    - Octonions
    - Sedenions
- Virasoro algebra

Generalities

- Algebraic structure
- Universal algebra
    - Variety (universal algebra)
    - Congruence relation
    - Free object
    - Generating set (universal algebra)
    - Clone (algebra)
- Kernel of a function
    - Kernel (algebra)
    - Isomorphism class
    - Isomorphism theorem
    - Fundamental theorem on homomorphisms
- Universal property
- Filtration (mathematics)
- Category theory
    - Monoidal category
    - Groupoid

    - Group object
    - Coalgebra
    - Bialgebra
    - Hopf algebra
    - Magma object
- Torsion (algebra)

COMPUTER ALGEBRA

- Symbolic mathematics
    - Finite field arithmetic
    - Gröbner basis
    - Buchberger's algorithm

OUTLINE OF ALGEBRAIC STRUCTURES

The following outline is provided as an overview of and topical guide to algebraic structures:

In mathematics, there are many types of algebraic structures which are studied. Abstract algebra is primarily the study of specific algebraic structures and their properties. Algebraic structures may be viewed in different ways, however the common starting point of algebra texts is that an algebraic object incorporates one or more sets with one or more binary operations or unary operations satisfying a collection of axioms.

Another branch of mathematics known as universal algebra studies algebraic structures in general. From the universal algebra viewpoint, most structures can be divided into varieties and quasivarieties depending on the axioms used. Some axiomatic formal systems that are neither varieties nor quasivarieties, called nonvarieties, are sometimes included among the algebraic structures by tradition.

Concrete examples of each structure will be found in the articles listed.

Algebraic structures are so numerous today that this article will inevitably be incomplete. In addition to this, there are sometimes multiple names for the same structure, and sometimes one name will be defined by disagreeing axioms by different authors. Most structures appearing on this page will be common ones which most authors agree on. Other web lists of algebraic structures, organized more or less alphabetically, include Jipsen and PlanetMath. These lists mention many structures not included below, and may present more information about some structures than is presented here.

Study of Algebraic Structures

Algebraic structures appear in most branches of mathematics, and one can encounter them in many different ways.

- Beginning study: In American universities, groups, vector spaces and fields are generally the first structures encountered in subjects such as linear algebra. They are usually introduced as sets with certain axioms.
- Advanced study:
    - Abstract algebra studies properties of specific algebraic structures.
    - Universal algebra studies algebraic structures abstractly, rather than specific types of structures.
        - Varieties
    - Category theory studies interrelationships between different structures, algebraic and non-algebraic. To study a non-algebraic object, it is often useful to use category theory to relate the object to an algebraic structure.
        - Example: The fundamental group of a topological space gives information about the topological space.

Types of Algebraic Structures

In full generality, an algebraic structure may use any number of sets and any number of axioms in its definition. The most commonly studied structures, however, usually involve only one or two sets and one or two binary operations. The structures below are organized by how many sets are involved, and how many binary operations are used. Increased indentation is meant to indicate a more exotic structure, and the least indented levels are the most basic.

One Binary Operation on One Set

The following structures consist of a set with a binary operation. The most common structure is that of a *group*. Other structures involve weakening or strengthening the axioms for groups, and may additionally use unary operations.

- Groups are key structures. Abelian groups are an important special type of group.
- semigroups and monoids: These are like groups, except the operation need not have inverse elements.
- quasigroups and loops: These are like groups, except the operation need not be associative.
- Magmas: These are like groups, except the operation need not be associative or have inverse elements.

- Semilattice: This is basically "half" of a lattice structure (see below).

Two Binary Operations on One Set

The main types of structures with one set having two binary operations are *rings* and *lattices*. The axioms defining many of the other structures are modifications of the axioms for rings and lattices. One major difference between rings and lattices is that their two operations are related to each other in different ways. In ring-like structures, the two operations are linked by the distributive law; in lattice-like structures, the operations are linked by the absorption law.

- Rings: The two operations are usually called addition and multiplication. Commutative rings are an especially important type of ring where the multiplication operation is commutative. Integral domains and fields are especially important types of commutative rings.
    - Nonassociative rings: These are like rings, but the multiplication operation need not be associative.
        - Lie rings and Jordan rings are special examples of nonassociative rings.
    - semirings: These are like rings, but the addition operation need not have inverses.
    - nearrings: These are like rings, but the addition operation need not be commutative.
    - *-rings: These are rings with an additional unary operation known as an involution.
- Lattices: The two operations are usually called meet and join.
    - Latticoid: meet and join commute but need not associate.
    - Skew lattice: meet and join associate but need not commute.

Two Binary Operations and two Sets

The following structures have the common feature of having two sets, A and B, so that there is a binary operation from $A \times A$ into A and another operation from $A \times B$ into A.

- Vector spaces: The set A is an abelian group, and the set B is a field.
    - Graded vector spaces: Vector spaces which are equipped with a direct sum decomposition into subspaces.
- Modules: The set A is an abelian group, but the B is only a general ring and not necessarily a field.

- Special types of modules, including free modules, projective modules, injective modules and flat modules are studied in abstract algebra.

■ Group with operators: In this case, the set *A* is a group, and the set *B* is just a set.

REFERENCES

1. Garrett Birkhoff, 1967. Lattice Theory, 3rd ed, AMS Colloquium Publications Vol. 25. American Mathematical Society.
2. ——, and Saunders MacLane, 1999 (1967). Algebra, 2nd ed. New York: Chelsea.
3. George Boolos and Richard Jeffrey, 1980. Computability and Logic, 2nd ed. Cambridge Univ. Press.
4. Dummit, David S., and Foote, Richard M., 2004. Abstract Algebra, 3rd ed. John Wiley and Sons.
5. Grätzer, George, 1978. Universal Algebra, 2nd ed. Springer.
6. David K. Lewis, 1991. Part of Classes. Blackwell.
7. Michel, Anthony N., and Herget, Charles J., 1993 (1981). Applied Algebra and Functional Analysis. Dover.
8. Potter, Michael, 2004. Set Theory and its Philosophy, 2nd ed. Oxford Univ. Press.
9. Smorynski, Craig, 1991. Logical Number Theory I. Springer-Verlag.
10. A monograph available free online:
11. Burris, Stanley N., and H.P. Sankappanavar, H. P., 1981. A Course in Universal Algebra. Springer-Verlag. ISBN 3-540-90578-2.

CHAPTER 11

Algebraic Structures

Chapter Contents

Abstract algebra is the study of algebraic structures. Such a structure consists of a set together with one or more binary operations, which are required to satisfy certain axioms. For example, here is the definition of a simple algebraic structure known as a group:

INTRODUCTION

In mathematics, a **group** is an algebraic structure consisting of a set of elements equipped with an operation that combines any two elements to form a third element. The operation satisfies four conditions called the group axioms, namely closure, associativity, identity and invertibility. One of the most familiar examples of a group is the set of integers together with the addition operation, but the abstract formalization of the group axioms, detached as it is from the concrete nature of any particular group and its operation, applies much more widely. It allows entities with highly diverse mathematical origins in abstract algebra and beyond to be handled in a flexible way while retaining their essential structural aspects. The ubiquity of groups in numerous areas within and outside mathematics makes them a central organizing principle of contemporary mathematics.

Groups share a fundamental kinship with the notion of symmetry. For example, a symmetry group encodes symmetry features of a geometrical object: the group consists of the set of transformations that leave the object unchanged and the operation of combining two such transformations by performing one after the other. Lie groups are the symmetry groups used in the Standard Model of particle physics; Poincaré groups, which are also Lie groups, can express the physical symmetry underlying special relativity; and point groups are used to help understand symmetry phenomena in molecular chemistry.

The concept of a group arose from the study of polynomial equations, starting with Évariste Galois in the 1830s. After contributions from other fields such as number theory and geometry, the group notion was generalized and firmly established around 1870. Modern group theory—an active mathematical discipline—studies groups in their own right.[a] To explore groups, mathematicians have devised various notions to break groups into smaller, better-understandable pieces, such as subgroups, quotient groups and simple groups. In addition to their abstract properties, group theorists also study the different ways in which a group can be expressed concretely, both from a point of view of representation theory (that is, through the representations of the group) and of computational group theory. A theory has been developed for finite groups, which culminated with the classification of finite simple groups, completed in 2004.[aa] Since the mid-1980s, geometric group theory, which studies finitely generated groups as geometric objects, has become a particularly active area in group theory.

The manipulations of this Rubik's Cube form the Rubik's Cube group.

DEFINITION AND ILLUSTRATION

First example: the integers

One of the most familiar groups is the set of integers **Z** which consists of the numbers ..., −4, −3, −2, −1, 0, 1, 2, 3, 4, ..., together with addition.

The following properties of integer addition serve as a model for the abstract group axioms given in the definition below.

- For any two integers a and b, the sum a + b is also an integer. That is, addition of integers always yields an integer. This property is known as closure under addition.
- For all integers a, b and c, (a + b) + c = a + (b + c). Expressed in words, adding a to b first, and then adding the result to c gives the same final result as adding a to the sum of b and c, a property known as associativity.
- If a is any integer, then 0 + a = a + 0 = a. Zero is called the identity element of addition because adding it to any integer returns the same integer.
- For every integer a, there is an integer b such that a + b = b + a = 0. The integer b is called the inverse element of the integer a and is denoted −a.

The integers, together with the operation +, form a mathematical object belonging to a broad class sharing similar structural aspects. To appropriately understand these structures as a collective, the following abstract definition is developed.

Definition

A group is a set, G, together with an operation • (called the group law of G) that combines any two elements a and b to form another element, denoted a • b or ab. To qualify as a group, the set and operation, (G, •), must satisfy four requirements known as the group axioms:

Closure

For all a, b in G, the result of the operation, a • b, is also in G.[b]

Associativity

For all a, b and c in G, (a • b) • c = a • (b • c).

Identity element

There exists an element e in G, such that for every element a in G, the equation e • a = a • e = a holds. Such an element is unique (see below), and thus one speaks of the identity element.

Inverse element

For each a in G, there exists an element b in G, commonly denoted a^{-1} (or −a, if the operation is denoted "+"), such that a • b = b • a = e, where e is the identity element.

The result of an operation may depend on the order of the operands. In other words, the result of combining element a with element b need not yield the same result as combining element b with element a; the equation a • b = b • a may not always be true. This equation always holds in the group of integers under addition, because a + b = b + a for any two integers (commutativity of addition). Groups for which the commutativity equation a • b = b • a always holds are called abelian groups (in honor of Niels Henrik Abel). The symmetry group described in the following section is an example of a group that is not abelian.

The identity element of a group G is often written as 1 or 1_G, a notation inherited from the multiplicative identity. If a group is abelian, then one may choose to denote the group operation by + and the identity element by 0; in that case, the group is called an additive group. The identity element can also be written as id.

The set G is called the underlying set of the group (G, •). Often the group's underlying set G is used as a short name for the group (G, •). Along the same lines, shorthand expressions such as "a subset of the group G" or "an element of group G" are used when what is actually meant is "a subset of the underlying set G of the group (G, •)" or "an element of the underlying set G of the group (G, •)". Usually, it is clear from the context whether a symbol like G refers to a group or to an underlying set.

The elements of the symmetry group of the square (D_4). Vertices are identified by color or number.

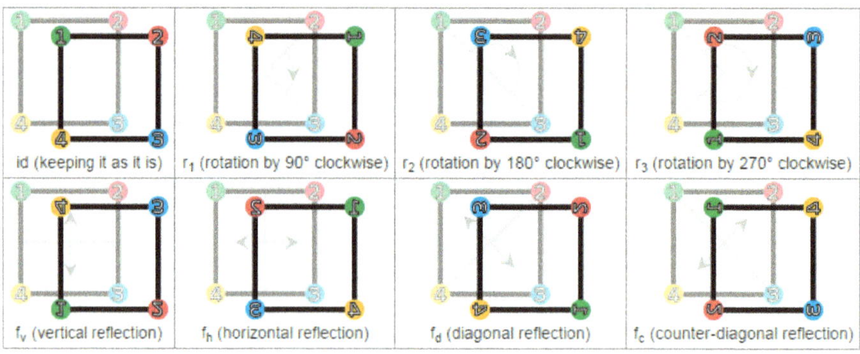

- the identity operation leaving everything unchanged, denoted id;
- rotations of the square around its center by 90° clockwise, 180° clockwise, and 270° clockwise, denoted by r_1, r_2 and r_3, respectively;
- reflections about the vertical and horizontal middle line (f_h and f_v), or through the two diagonals (f_d and f_c).

These symmetries are represented by functions. Each of these functions sends a point in the square to the corresponding point under the symmetry. For example, r_1 sends a point to its rotation 90° clockwise around the square's center, and f_h sends a point to its reflection across the square's vertical middle line. Composing two of these symmetry functions gives another symmetry function. These symmetries determine a group called the dihedral group of degree 4 and denoted D_4. The underlying set of the group is the above set of symmetry functions, and the group operation is function composition. Two symmetries are combined by composing them as functions, that is, applying the first one to the square, and the second one to the result of the first application. The result of performing first a and then b is written symbolically from right to left as b • a ("apply the symmetry b after performing the symmetry a").

The right-to-left notation is the same notation that is used for composition of functions.

The group table on the right lists the results of all such compositions possible. For example, rotating by 270° clockwise (r_3) and then reflecting horizontally (f_h) is the same as performing a reflection along the diagonal (f_d). Using the above symbols, highlighted in blue in the group table:

$f_h \bullet r_3 = f_d$.

Group table of D_4

•	id	r_1	r_2	r_3	f_v	f_h	f_d	f_c
id	id	r_1	r_2	r_3	f_v	f_h	f_d	f_c
r_1	r_1	r_2	r_3	id	f_c	f_d	f_v	f_h
r_2	r_2	r_3	id	r_1	f_h	f_v	f_c	f_d
r_3	r_3	id	r_1	r_2	f_d	f_c	f_h	f_v
f_v	f_v	f_d	f_h	f_c	id	r_2	r_1	r_3
f_h	f_h	f_c	f_v	f_d	r_2	id	r_3	r_1
f_d	f_d	f_h	f_c	f_v	r_3	r_1	id	r_2
f_c	f_c	f_v	f_d	f_h	r_1	r_3	r_2	id

The elements id, r_1, r_2, and r_3 form a subgroup, highlighted in ☐ red (upper left region). A left and right coset of this subgroup is highlighted in ☐ green (in the last row) and ☐ yellow (last column), respectively.

Given this set of symmetries and the described operation, the group axioms can be understood as follows:

- The closure axiom demands that the composition b • a of any two symmetries a and b is also a symmetry. Another example for the group operation is

$r_3 \bullet f_h = f_c$,

i.e. rotating 270° clockwise after reflecting horizontally equals reflecting along the counter-diagonal (f_c). Indeed every other combination of two symmetries still gives a symmetry, as can be checked using the group table.

- The associativity constraint deals with composing more than two symmetries: Starting with three elements a, b and c of D_4, there are two possible ways of using these three symmetries in this order to determine a symmetry of the square. One of these ways is to first compose a and b into a single symmetry, then to compose that symmetry with c. The other way is to first compose b and c, then to compose the resulting symmetry with a. The associativity condition

$(a \bullet b) \bullet c = a \bullet (b \bullet c)$

means that these two ways are the same, i.e., a product of many group elements can be simplified in any grouping. For example, $(f_d \bullet f_v) \bullet r_2 = f_d \bullet (f_v \bullet r_2)$ can be checked using the group table at the right

$(f_d \bullet f_v) \bullet r_2$	=	$r_3 \bullet r_2$	=	r_1, which equals
$f_d \bullet (f_v \bullet r_2)$	=	$f_d \bullet f_h$	=	r_1.

\While associativity is true for the symmetries of the square and addition of numbers, it is not true for all operations. For instance, subtraction of numbers is not associative: $(7 - 3) - 2 = 2$ is not the same as $7 - (3 - 2) = 6$.

- The identity element is the symmetry id leaving everything unchanged: for any symmetry a, performing id after a (or a after id) equals a, in symbolic form,

$\text{id} \bullet a = a$,

$a \bullet \text{id} = a$.

- An inverse element undoes the transformation of some other element. Every symmetry can be undone: each of the following transformations—identity id, the reflections f_h, f_v, f_d, f_c and the 180° rotation r_2—is its own inverse, because performing it twice brings the square back to its original orientation. The rotations r_3 and r_1 are each other's inverses, because rotating 90° and then rotation 270° (or vice versa) yields a rotation over 360° which leaves the square unchanged. In symbols,

$f_h \bullet f_h = \text{id}$,

$r_3 \bullet r_1 = r_1 \bullet r_3 = \text{id}$.

In contrast to the group of integers above, where the order of the operation is irrelevant, it does matter in D_4: $f_h \bullet r_1 = f_c$ but $r_1 \bullet f_h = f_d$. In other words, D_4 is not abelian, which makes the group structure more difficult than the integers introduced first.

HISTORY

The modern concept of an abstract group developed out of several fields of mathematics. The original motivation for group theory was the quest for solutions of polynomial equations of degree higher than 4. The 19th-century French mathematician Évariste Galois, extending prior work of Paolo Ruffini and Joseph-Louis Lagrange, gave a criterion for the solvability of a particular polynomial equation in terms of the symmetry group of its roots (solutions). The elements of such a Galois group correspond to certain permutations of the roots. At first, Galois' ideas were rejected by his contemporaries, and published only posthumously. More general permutation groups were investigated in particular by Augustin Louis Cauchy. Arthur Cayley's On the theory of groups, as depending on the symbolic equation $\theta^n = 1$ (1854) gives the first abstract definition of a finite group.

Geometry was a second field in which groups were used systematically, especially symmetry groups as part of Felix Klein's 1872 Erlangen program. After novel geometries such as hyperbolic and projective geometry had emerged, Klein used group theory to organize them in a more coherent way. Further advancing these ideas, Sophus Lie founded the study of Lie groups in 1884.

The third field contributing to group theory was number theory. Certain abelian group structures had been used implicitly in Carl Friedrich Gauss' number-theoretical work Disquisitiones Arithmeticae (1798), and more explicitly by Leopold Kronecker. In 1847, Ernst Kummer made early attempts to prove Fermat's Last Theorem by developing groups describing factorization into prime numbers.

The convergence of these various sources into a uniform theory of groups started with Camille Jordan's Traité des substitutions et des équations algébriques (1870). Walther von Dyck (1882) introduced the idea of specifying a group by means of generators and relations, and was also the first to give an axiomatic definition of an "abstract group", in the terminology of the time. As of the 20th century, groups gained wide recognition by the pioneering work of Ferdinand Georg Frobenius and William Burnside, who worked on representation theory of finite groups, Richard Brauer's modular representation theory and Issai Schur's papers. The theory of Lie groups, and more generally locally compact groups was studied by Hermann Weyl, Élie Cartan and many others. Its algebraic counterpart, the theory of algebraic groups, was first shaped by Claude Chevalley (from the late 1930s) and later by the work of Armand Borel and Jacques Tits.

The University of Chicago's 1960–61 Group Theory Year brought together group theorists such as Daniel Gorenstein, John G. Thompson and Walter Feit, laying the foundation of a collaboration that, with input from numerous other mathematicians, led to the classification of finite simple groups, with the final step taken by Aschbacher and Smith in 2004. This project exceeded previous mathematical endeavours by its sheer size, in both length of proof and number of researchers. Research is ongoing to simplify the proof of this classification. These days, group theory is still a highly active mathematical branch, impacting many other fields.[a]

ELEMENTARY CONSEQUENCES OF THE GROUP AXIOMS

Basic facts about all groups that can be obtained directly from the group axioms are commonly subsumed under elementary group theory. For example, repeated applications of the associativity axiom show that the unambiguity of a • b • c = (a • b) • c = a • (b • c) generalizes to more than three factors. Because this implies that parentheses can be inserted anywhere within such a series of terms, parentheses are usually omitted.

The axioms may be weakened to assert only the existence of a left identity and left inverses. Both can be shown to be actually two-sided, so the resulting definition is equivalent to the one given above.

Uniqueness of identity element and inverses

Two important consequences of the group axioms are the uniqueness of the identity element and the uniqueness of inverse elements. There can be only one identity element in a group, and each element in a group has exactly one inverse element. Thus, it is customary to speak of the identity, and the inverse of an element.

To prove the uniqueness of an inverse element of a, suppose that a has two inverses, denoted b and c, in a group (G, •). Then

$b = b \cdot e$	as e is the identity element
$= b \cdot (a \cdot c)$	because c is an inverse of a, so $e = a \cdot c$
$= (b \cdot a) \cdot c$	by associativity, which allows to rearrange the parentheses
$= e \cdot c$	since b is an inverse of a, i.e. $b \cdot a = e$
$= c$	for e is the identity element

The term b on the first line above and the c on the last are equal, since they are connected by a chain of equalities. In other words, there is only one inverse element of a. Similarly, to prove that the identity element of a group is unique, assume G is a group with two identity elements e and f. Then e = e • f = f, hence e and f are equal.

Division

In groups, the existence of inverse elements implies that division is possible: given elements a and b of the group G, there is exactly one solution x in G to the equation x • a = b, namely $b \cdot a^{-1}$. In fact, we have

$(b \cdot a^{-1}) \cdot a = b \cdot (a^{-1} \cdot a) = b \cdot e = b.$

Uniqueness results by multiplying the two sides of the equation x • a = b by a^{-1}. The element $b \cdot a^{-1}$, often denoted b / a, is called the right quotient of b by a, or the result of the right division of b by a.

Similarly there is exactly one solution y in G to the equation a • y = b, namely $y = a^{-1} \cdot b$. This solution is the left quotient of b by a, and is sometimes denoted a \ b.

In general a / b and a \ b may be different, but, if the group operation is commutative (that is, if the group is abelian), they are equal. In this case, the group operation is often denoted as an addition, and one talks of subtraction and difference instead of division and quotient.

A consequence of this is that multiplication by a group element g is a bijection. Specifically, if g is an element of the group G, the function (mathematics) from G to itself that maps h ∈ G to g • h is a bijection. This function is called the left translation by g . Similarly, the right translation by g is the bijection from G to itself, that maps h to h • g. If G is abelian, the left and the right translation by a group element are the same.

BASIC CONCEPTS

The following sections use mathematical symbols such as X = {x, y, z} to denote a set X containing elements x, y, and z, or alternatively x ∈ X to restate that x is an element of X. The notation f : X → Y means f is a function assigning to every element of X an element of Y.

Further information: Glossary of group theory

To understand groups beyond the level of mere symbolic manipulations as above, more structural concepts have to be employed.[c] There is a conceptual principle underlying all of the following notions: to take advantage of the structure offered by groups (which sets, being "structureless", do not have), constructions related to groups have to be compatible with the group operation. This compatibility manifests itself in the following notions in various ways. For example, groups can be related to each other via functions called group homomorphisms. By the mentioned principle, they are required to respect the group structures in a precise sense. The structure of groups can also be understood by breaking them into pieces called subgroups and quotient groups. The principle of "preserving structures"—a recurring topic in mathematics throughout—is an instance of working in a category, in this case the category of groups.

Group homomorphisms

Group homomorphisms[g] are functions that preserve group structure. A function $a: G \to H$ between two groups $(G, \bullet)$ and $(H, *)$ is called a homomorphism if the equation

$a(g \bullet k) = a(g) * a(k)$ holds for all elements g, k in G. In other words, the result is the same when performing the group operation after or before applying the map a. This requirement ensures that $a(1_G) = 1_H$, and also $a(g)^{-1} = a(g^{-1})$ for all g in G. Thus a group homomorphism respects all the structure of G provided by the group axioms.

Two groups G and H are called isomorphic if there exist group homomorphisms $a: G \to H$ and $b: H \to G$, such that applying the two functions one after another in each of the two possible orders gives the identity functions of G and H. That is, $a(b(h)) = h$ and $b(a(g)) = g$ for any g in G and h in H. From an abstract point of view, isomorphic groups carry the same information. For example, proving that $g \bullet g = 1_G$ for some element g of G is equivalent to proving that $a(g) * a(g) = 1_H$, because applying a to the first equality yields the second, and applying b to the second gives back the first.

Subgroups

Informally, a subgroup is a group H contained within a bigger one, G. Concretely, the identity element of G is contained in H, and whenever h_1 and h_2 are in H, then so are $h_1 \bullet h_2$ and h_1^{-1}, so the elements of H, equipped with the group operation on G restricted to H, indeed form a group.

In the example above, the identity and the rotations constitute a subgroup $R = \{id, r_1, r_2, r_3\}$, highlighted in red in the group table above: any two rotations composed are still a rotation, and a rotation can be undone by (i.e. is inverse to) the complementary rotations 270° for 90°, 180° for 180°, and 90° for 270° (note that rotation in the opposite direction is not defined). The subgroup test is a necessary and sufficient condition for a nonempty subset H of a group G to be a subgroup: it is sufficient to check that $g^{-1}h \in H$ for all elements $g, h \in H$. Knowing the subgroups is important in understanding the group as a whole.[d]

Given any subset S of a group G, the subgroup generated by S consists of products of elements of S and their inverses. It is the smallest subgroup of G containing S. In the introductory example above, the subgroup generated by r_2 and f_v consists of these two elements, the identity element id and $f_h = f_v \cdot r_2$. Again, this is a subgroup, because combining any two of these four elements or their inverses (which are, in this particular case, these same elements) yields an element of this subgroup.

Cosets

In many situations it is desirable to consider two group elements the same if they differ by an element of a given subgroup. For example, in D_4 above, once a reflection is performed, the square never gets back to the r_2 configuration by just applying the rotation operations (and no further reflections), i.e. the rotation operations are irrelevant to the question whether a reflection has been performed. Cosets are used to formalize this insight: a subgroup H defines left and right cosets, which can be thought of as translations of H by arbitrary group elements g. In symbolic terms, the left and right cosets of H containing g are

$gH = \{g \cdot h : h \in H\}$ and $Hg = \{h \cdot g : h \in H\}$, respectively.

The left cosets of any subgroup H form a partition of G; that is, the union of all left cosets is equal to G and two left cosets are either equal or have an emptyintersection. The first case $g_1H = g_2H$ happens precisely when $g_1^{-1} \cdot g_2 \in H$, i.e. if the two elements differ by an element of H. Similar considerations apply to the right cosets of H. The left and right cosets of H may or may not be equal. If they are, i.e. for all g in G, $gH = Hg$, then H is said to be a normal subgroup.

In D_4, the introductory symmetry group, the left cosets gR of the subgroup R consisting of the rotations are either equal to R, if g is an element of R itself, or otherwise equal to $U = f_cR = \{f_c, f_v, f_d, f_h\}$ (highlighted in green). The subgroup R is also normal, because $f_cR = U = Rf_c$ and similarly for any element other than f_c. (In fact, in the case of D_4, observe that all such cosets are equal, such that $f_hR = f_vR = f_dR = f_cR$.)

Quotient groups

In some situations the set of cosets of a subgroup can be endowed with a group law, giving a quotient group or factor group. For this to be possible, the subgroup has to be normal. Given any normal subgroup N, the quotient group is defined by

$G / N = \{gN, g \in G\}$, "G modulo N".

This set inherits a group operation (sometimes called coset multiplication, or coset addition) from the original group G: $(gN) \cdot (hN) = (gh)N$ for all g and h in G. This definition is motivated by the idea (itself an instance of general structural considerations outlined above) that the map $G \to G / N$ that associates to any element g its coset gN be a group homomorphism, or by general abstract considerations called universal properties. The coset $eN = N$ serves as the identity in this group, and the inverse of gN in the quotient group is $(gN)^{-1} = (g^{-1})N$.[e]

The elements of the quotient group D_4 / R are R itself, which represents the identity, and U = f_vR. The group operation on the quotient is shown at the right. For example, U • U = f_vR • f_vR = (f_v • f_v)R = R. Both the subgroup R = {id, r_1, r_2, r_3}, as well as the corresponding quotient are abelian, whereas D_4 is not abelian. Building bigger groups by smaller ones, such as D_4 from its subgroup R and the quotient D_4 / R is abstracted by a notion called semidirect product.

Group table of the quotient group D_4 / R

•	**R**	**U**
R	R	U
U	U	R

Quotient groups and subgroups together form a way of describing every group by its presentation: any group is the quotient of the free group over the generators of the group, quotiented by the subgroup of relations. The dihedral group D_4, for example, can be generated by two elements r and f (for example, r = r_1, the right rotation and f = f_v the vertical (or any other) reflection), which means that every symmetry of the square is a finite composition of these two symmetries or their inverses. Together with the relations

$r^4 = f^2 = (r \cdot f)^2 = 1$,

the group is completely described. A presentation of a group can also be used to construct the Cayley graph, a device used to graphically capture discrete groups.

Sub- and quotient groups are related in the following way: a subset H of G can be seen as an injective map H → G, i.e. any element of the target has at most one element that maps to it. The counterpart to injective maps are surjective maps (every element of the target is mapped onto), such as the canonical map G → G / N.[y] Interpreting subgroup and quotients in light of these homomorphisms emphasizes the structural concept inherent to these definitions alluded to in the introduction. In general, homomorphisms are neither injective nor surjective. Kernel and image of group homomorphisms and the first isomorphism theorem address this phenomenon.

EXAMPLES AND APPLICATIONS

Examples and applications of groups abound. A starting point is the group **Z** of integers with addition as group operation, introduced above. If instead of addition multiplication is considered, one obtains multiplicative groups. These groups are predecessors of important constructions in abstract algebra.

Groups are also applied in many other mathematical areas. Mathematical objects are often examined by associating groups to them and studying the properties of the corresponding groups. For example, Henri Poincaré founded what is now called algebraic topology by introducing the fundamental group. By means of this connection, topological properties such as proximity and continuity translate into properties of groups.[i] For example, elements of the fundamental group are represented by loops. The second image at the right shows some loops in a plane minus a point. The blue loop is considered null-homotopic (and thus irrelevant), because it can be continuously shrunk to a point. The presence of the hole prevents the orange loop from being shrunk to a point. The fundamental group of the plane with a point deleted turns out to be infinite cyclic, generated by the orange loop (or any other loop winding once around the hole). This way, the fundamental group detects the hole.

In more recent applications, the influence has also been reversed to motivate geometric constructions by a group-theoretical background.[j] In a similar vein, geometric group theory employs geometric concepts, for example in the study of hyperbolic groups. Further branches crucially applying groups include algebraic geometry and number theory.

A periodic wallpaper pattern gives rise to a wallpaper group.

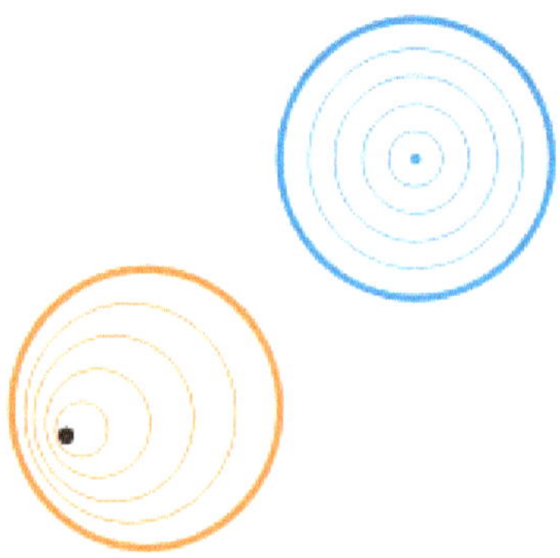

The fundamental group of a plane minus a point (bold) consists of loops around the missing point. This group is isomorphic to the integers.

In addition to the above theoretical applications, many practical applications of groups exist. Cryptography relies on the combination of the abstract group theory approach together with algorithmical knowledge obtained in computational group theory, in particular when implemented for finite groups. Applications of group theory are not restricted to mathematics; sciences such as physics, chemistry and computer science benefit from the concept.

Numbers

Many number systems, such as the integers and the rationals enjoy a naturally given group structure. In some cases, such as with the rationals, both addition and multiplication operations give rise to group structures. Such number systems are predecessors to more general algebraic structures known as rings and fields. Further abstract algebraic concepts such as modules, vector spaces and algebras also form groups.

Integers

The group of integers **Z** under addition, denoted (**Z**, +), has been described above. The integers, with the operation of multiplication instead of addition, (**Z**, ·) do not form a group. The closure, associativity and identity axioms are satisfied, but inverses do not exist: for example, a = 2 is an integer, but the only solution to the equation a · b = 1 in this case is b = 1/2, which is a rational number, but not an integer. Hence not every element of **Z** has a (multiplicative) inverse.[k]

Rationals

The desire for the existence of multiplicative inverses suggests considering fractions

$$\frac{a}{b}.$$

Fractions of integers (with b nonzero) are known as rational numbers.[l] The set of all such fractions is commonly denoted **Q**. There is still a minor obstacle for (**Q**, ·), the rationals with multiplication, being a group: because the rational number 0 does not have a multiplicative inverse (i.e., there is no x such that x · 0 = 1), (**Q**, ·) is still not a group.

However, the set of all nonzero rational numbers $\mathbf{Q} \setminus \{0\} = \{q \in \mathbf{Q} \mid q \neq 0\}$ does form an abelian group under multiplication, denoted ($\mathbf{Q} \setminus \{0\}$, ·).[m] Associativity and identity element axioms follow from the properties of integers. The closure requirement still holds true after removing zero, because the product of two nonzero rationals is never zero. Finally, the inverse of a/b is b/a, therefore the axiom of the inverse element is satisfied.

The rational numbers (including 0) also form a group under addition. Intertwining addition and multiplication operations yields more complicated structures called rings and—if division is possible, such as in **Q**—fields, which occupy a central position in abstract algebra. Group theoretic arguments therefore underlie parts of the theory of those entities.

Modular Arithmetic

In modular arithmetic, two integers are added and then the sum is divided by a positive integer called the modulus. The result of modular addition is the remainder of that division. For any modulus, n, the set of integers from 0 to n − 1 forms a group under modular addition: the inverse of any element a is n − a, and 0 is the identity element.

This is familiar from the addition of hours on the face of a clock: if the hour hand is on 9 and is advanced 4 hours, it ends up on 1, as shown at the right. This is expressed by saying that 9 + 4 equals 1 "modulo 12" or, in symbols,

$9 + 4 \equiv 1$ modulo 12.

The group of integers modulo n is written $\mathbf{Z}_n$ or $\mathbf{Z}/n\mathbf{Z}$.

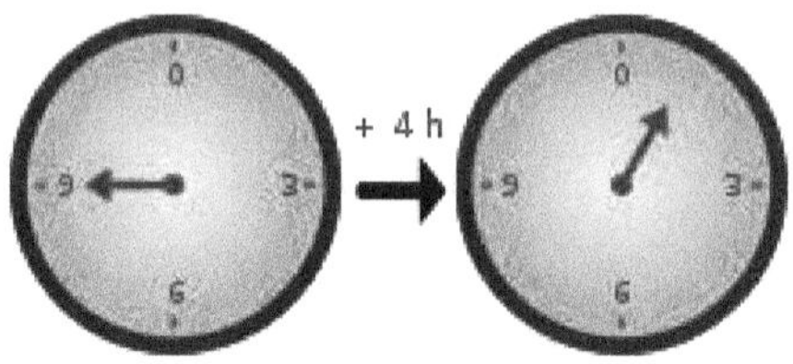

The hours on a clock form a group that uses addition modulo 12. Here 9 + 4 = 1.

For any prime number p, there is also the multiplicative group of integers modulo p. Its elements are the integers 1 to p − 1. The group operation is multiplication modulo p. That is, the usual product is divided by p and the remainder of this division is the result of modular multiplication. For example, if p = 5, there are four group elements 1, 2, 3, 4. In this group, $4 \cdot 4 = 1$, because the usual product 16 is equivalent to 1, which divided by 5 yields a remainder of 1. for 5 divides 16 − 1 = 15, denoted

$16 \equiv 1 \pmod 5$.

The primality of p ensures that the product of two integers neither of which is divisible by p is not divisible by p either, hence the indicated set of classes is closed under multiplication.[o] The identity element is 1, as usual for a multiplicative group, and the associativity follows from the corresponding property of integers. Finally, the inverse element axiom requires that given an integer a not divisible by p, there exists an integer b such that

$a \cdot b \equiv 1 \pmod p$, i.e. p divides the difference $a \cdot b - 1$.

The inverse b can be found by using Bézout's identity and the fact that the greatest common divisor gcd(a, p) equals 1. In the case p = 5 above, the inverse of 4 is 4, and the inverse of 3 is 2, as $3 \cdot 2 = 6 \equiv 1 \pmod 5$. Hence all group axioms are fulfilled. Actually, this example is similar to $(\mathbf{Q} \setminus \{0\}, \cdot)$ above: it consists of exactly those elements in $\mathbf{Z}/p\mathbf{Z}$ that have a multiplicative inverse. These groups are denoted $\mathbf{F}_p^\times$. They are crucial to public-key cryptography.[p]

Cyclic Groups

A cyclic group is a group all of whose elements are powers of a particular element a. In multiplicative notation, the elements of the group are:

$\ldots, a^{-3}, a^{-2}, a^{-1}, a^0 = e, a, a^2, a^3, \ldots,$

where a^2 means $a \bullet a$, and a^{-3} stands for $a^{-1} \bullet a^{-1} \bullet a^{-1} = (a \bullet a \bullet a)^{-1}$ etc.[h] Such an element a is called a generator or a primitive element of the group. In additive notation, the requirement for an element to be primitive is that each element of the group can be written as

..., −a−a, −a, 0, a, a+a, ...

In the groups **Z**/n**Z** introduced above, the element 1 is primitive, so these groups are cyclic. Indeed, each element is expressible as a sum all of whose terms are 1. Any cyclic group with n elements is isomorphic to this group. A second example for cyclic groups is the group of n-th complex roots of unity, given by complex numbers z satisfying $z^n = 1$. These numbers can be visualized as the vertices on a regular n-gon, as shown in blue at the right for n = 6. The group operation is multiplication of complex numbers. In the picture, multiplying with z corresponds to a counter-clockwise rotation by 60°. Using some field theory, the group $\mathbf{F}_p^\times$ can be shown to be cyclic: for example, if p = 5, 3 is a generator since $3^1 = 3$, $3^2 = 9 \equiv 4$, $3^3 \equiv 2$, and $3^4 \equiv 1$.

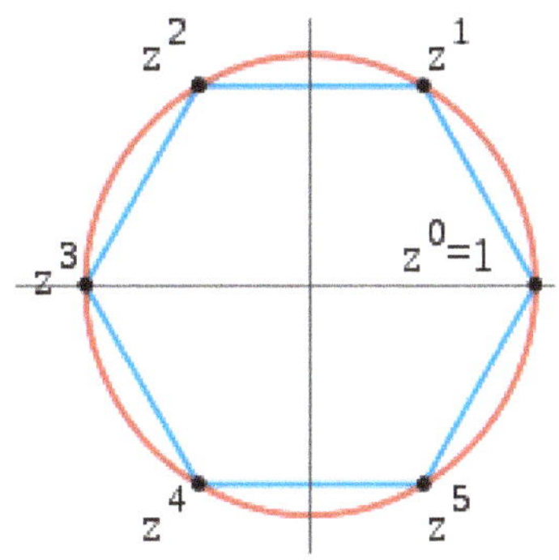

The 6th complex roots of unity form a cyclic group. z is a primitive element, but z^2 is not, because the odd powers of z are not a power of z^2.

Some cyclic groups have an infinite number of elements. In these groups, for every non-zero element a, all the powers of a are distinct; despite the name "cyclic group", the powers of the elements do not cycle. An infinite cyclic group is isomorphic to (**Z**, +), the group of integers under addition introduced above. As these two prototypes are both abelian, so is any cyclic group.

The study of finitely generated abelian groups is quite mature, including the fundamental theorem of finitely generated abelian groups; and reflecting this state of affairs, many group-related notions, such as center and commutator, describe the extent to which a given group is not abelian.

Symmetry Groups

Symmetry groups are groups consisting of symmetries of given mathematical objects—be they of geometric nature, such as the introductory symmetry group of the square, or of algebraic nature, such as polynomial equations and their solutions. Conceptually, group theory can be thought of as the study of symmetry.'Symmetries in mathematics greatly simplify the study of geometrical or analytical objects. A group is said to act on another mathematical object X if every group element performs some operation on X compatibly to the group law. In the rightmost example below, an element of order 7 of the (2,3,7) triangle group acts on the tiling by permuting the highlighted warped triangles (and the other ones, too). By a group action, the group pattern is connected to the structure of the object being acted on.

Rotations and reflections form the symmetry group of a great icosahedron.

In chemical fields, such as crystallography, space groups and point groups describe molecular symmetries and crystal symmetries. These symmetries underlie the chemical and physical behavior of these systems, and group theory enables simplification of quantum mechanical analysis of these properties. For example, group theory is used to show that optical transitions between certain quantum levels cannot occur simply because of the symmetry of the states involved.

Not only are groups useful to assess the implications of symmetries in molecules, but surprisingly they also predict that molecules sometimes can change symmetry. The Jahn-Teller effect is a distortion of a molecule of high symmetry when it adopts a particular ground state of lower symmetry from a set of possible ground states that are related to each other by the symmetry operations of the molecule.

Likewise, group theory helps predict the changes in physical properties that occur when a material undergoes a phase transition, for example, from a cubic to a tetrahedral crystalline form. An example is ferroelectric materials, where the change from a paraelectric to a ferroelectric state occurs at the Curie temperature and is related to a change from the high-symmetry paraelectric state to the lower symmetry ferroelectric state, accompanied by a so-called soft phonon mode, a vibrational lattice mode that goes to zero frequency at the transition.

Such spontaneous symmetry breaking has found further application in elementary particle physics, where its occurrence is related to the appearance of Goldstone bosons.

Finite symmetry groups such as the Mathieu groups are used in coding theory, which is in turn applied in error correction of transmitted data, and in CD players. Another application is differential Galois theory, which characterizes functions having antiderivatives of a prescribed form, giving group-theoretic criteria for when solutions of certain differential equations are well-behaved.[u] Geometric properties that remain stable under group actions are investigated in (geometric) invariant theory.

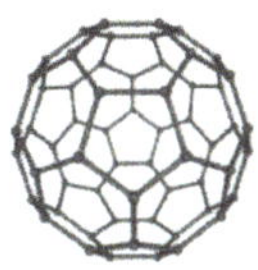			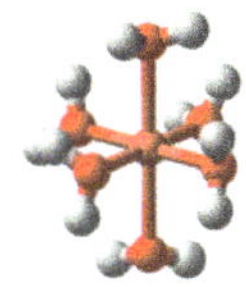	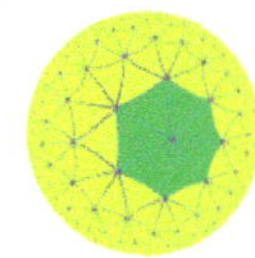
Buckminsterfullerene displays icosahedral symmetry, though the double bonds reduce this to pyritohedral symmetry.	Ammonia, NH_3. Its symmetry group is of order 6, generated by a 120° rotation and a reflection.	Cubane C_8H_8 features octahedral symmetry.	Hexaaquacopper(II) complex ion, $[Cu(OH_2)_6]^{2+}$. Compared to a perfectly symmetrical shape, the molecule is vertically dilated by about 22% (Jahn-Teller effect).	The (2,3,7) triangle group, a hyperbolic group, acts on this tiling of the hyperbolic plane.

General Linear Group and Representation Theory

Matrix groups consist of matrices together with matrix multiplication. The general linear group GL(n, **R**) consists of all invertible n-by-n matrices with real entries. Its subgroups are referred to as matrix groups or linear groups. The dihedral group example mentioned above can be viewed as a (very small) matrix group. Another important matrix group is the special orthogonal group SO(n). It describes all possible rotations in n dimensions. Via Euler angles, rotation matrices are used in computer graphics.

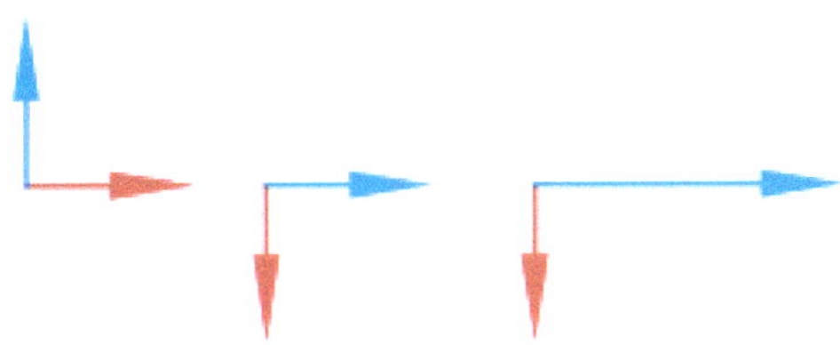

Two vectors (the left illustration) multiplied by matrices (the middle and right illustrations). The middle illustration represents a clockwise rotation by 90°, while the right-most one stretches the x-coordinate by factor 2.

Representation theory is both an application of the group concept and important for a deeper understanding of groups. It studies the group by its group actions on other spaces. A broad class of group representations are linear representations, i.e. the group is acting on a vector space, such as the three-dimensional Euclidean space $\mathbf{R}^3$. A representation of G on an n-dimensional real vector space is simply a group homomorphism

$\rho: G \to GL(n, \mathbf{R})$

from the group to the general linear group. This way, the group operation, which may be abstractly given, translates to the multiplication of matrices making it accessible to explicit computations.

Given a group action, this gives further means to study the object being acted on.[x] On the other hand, it also yields information about the group. Group representations are an organizing principle in the theory of finite groups, Lie groups, algebraic groups and topological groups, especially (locally) compact groups.

Galois Groups

Galois groups were developed to help solve polynomial equations by capturing their symmetry features. For example, the solutions of the quadratic equationax^2 + bx + c = 0 are given by

$$x = \frac{-b \pm \sqrt{b^2 - 4ac}}{2a}.$$

Exchanging "+" and "–" in the expression, i.e. permuting the two solutions of the equation can be viewed as a (very simple) group operation. Similar formulae are known for cubic and quartic equations, but do not exist in general for degree 5 and higher. Abstract properties of Galois groups associated with polynomials (in particular their solvability) give a criterion for polynomials that have all their solutions expressible by radicals, i.e. solutions expressible using solely addition, multiplication, and roots similar to the formula above.

The problem can be dealt with by shifting to field theory and considering the splitting field of a polynomial. Modern Galois theory generalizes the above type of Galois groups to field extensions and establishes—via the fundamental theorem of Galois theory—a precise relationship between fields and groups, underlining once again the ubiquity of groups in mathematics.

FINITE GROUPS

A group is called finite if it has a finite number of elements. The number of elements is called the order of the group. An important class is the symmetric groups S_N, the groups of permutations of N letters. For example, the symmetric group on 3 letters S_3 is the group consisting of all possible orderings of the three letters ABC, i.e. contains the elements ABC, ACB, BAC, BCA, CAB, CBA, in total 6 (factorial of 3) elements. This class is fundamental insofar as any finite group can be expressed as a subgroup of a symmetric group S_N for a suitable integer N, according to Cayley's theorem. Parallel to the group of symmetries of the square above, S_3 can also be interpreted as the group of symmetries of an equilateral triangle.

The order of an element a in a group G is the least positive integer n such that a^n = e, where a^n represents

$$\underbrace{a \cdots a}_{n \text{ factors}},$$

i.e. application of the operation • to n copies of a. (If • represents multiplication, then a^n corresponds to the nth power of a.) In infinite groups, such an n may not exist, in which case the order of a is said to be infinity. The order of an element equals the order of the cyclic subgroup generated by this element.

More sophisticated counting techniques, for example counting cosets, yield more precise statements about finite groups: Lagrange's Theorem states that for a finite group G the order of any finite subgroup H divides the order of G. The Sylow theorems give a partial converse.

The dihedral group (discussed above) is a finite group of order 8. The order of r_1 is 4, as is the order of the subgroup R it generates (see above). The order of the reflection elements f_v etc. is 2. Both orders divide 8, as predicted by Lagrange's theorem. The groups $\mathbf{F}_p^\times$ above have order p − 1.

Classification of Finite Simple Groups

Mathematicians often strive for a complete classification (or list) of a mathematical notion. In the context of finite groups, this aim leads to difficult mathematics. According to Lagrange's theorem, finite groups of order p, a prime number, are necessarily cyclic (abelian) groups $\mathbf{Z}_p$. Groups of order p^2 can also be shown to be abelian, a statement which does not generalize to order p^3, as the non-abelian group D_4 of order $8 = 2^3$ above shows. Computer algebra systems can be used to list small groups, but there is no classification of all finite groups.[q] An intermediate step is the classification of finite simple groups.[r] A nontrivial group is called simple if its only normal subgroups are the trivial group and the group itself.[s] The Jordan–Hölder theorem exhibits finite simple groups as the building blocks for all finite groups. Listing all finite simple groups was a major achievement in contemporary group theory. 1998 Fields Medal winner Richard Borcherds succeeded in proving the monstrous moonshine conjectures, a surprising and deep relation between the largest finite simple sporadic group—the "monster group"—and certain modular functions, a piece of classical complex analysis, and string theory, a theory supposed to unify the description of many physical phenomena.

GROUPS WITH ADDITIONAL STRUCTURE

Many groups are simultaneously groups and examples of other mathematical structures. In the language of category theory, they are group objects in a category, meaning that they are objects (that is, examples of another mathematical structure) which come with transformations (called morphisms) that mimic the group axioms. For example, every group (as defined above) is also a set, so a group is a group object in the category of sets.

Topological Groups

Some topological spaces may be endowed with a group law. In order for the group law and the topology to interweave well, the group operations must be continuous functions, that is, $g \bullet h$, and g^{-1} must not vary wildly if g and h vary only little. Such groups are called topological groups, and they are the group objects in the category of topological spaces. The most basic examples are the reals **R** under addition, ($\mathbf{R} \setminus \{0\}$, ·), and similarly with any other topological field such as the complex numbers or p-adic numbers. All of these groups are locally compact, so they have Haar measures and can be studied via harmonic analysis. The former offer an abstract formalism of invariant integrals. Invariance means, in the case of real numbers for example:

$$\int f(x)\,dx = \int f(x+c)\,dx$$

for any constant c. Matrix groups over these fields fall under this regime, as do adele rings and adelic algebraic groups, which are basic to number theory. Galois groups of infinite field extensions such as the absolute Galois group can also be equipped with a topology, the so-called Krull topology, which in turn is central to generalize the above sketched connection of fields and groups to infinite field extensions. An advanced generalization of this idea, adapted to the needs of algebraic geometry, is the étale fundamental group.

Lie Groups

Lie groups (in honor of Sophus Lie) are groups which also have a manifold structure, i.e. they are spaces looking locally like some Euclidean space of the appropriate dimension. Again, the additional structure, here the manifold structure, has to be compatible, i.e. the maps corresponding to multiplication and the inverse have to be smooth.

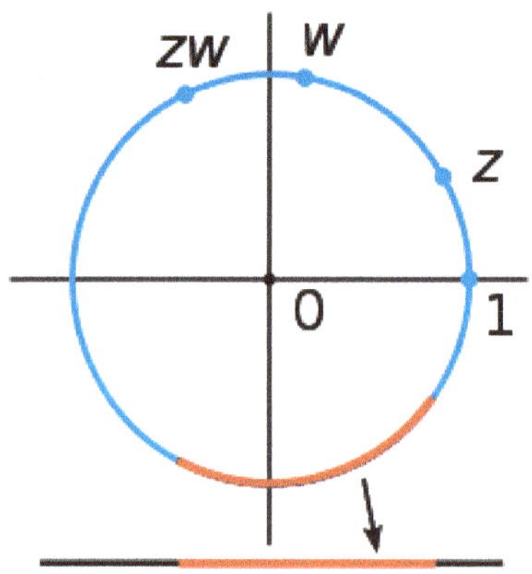

The unit circle in the complex plane under complex multiplication is a Lie group and, therefore, a topological group. It is topological since complex multiplication and division are continuous. It is a manifold and thus a Lie group, because every small piece, such as the red arc in the figure, looks like a part of the real line (shown at the bottom).

A standard example is the general linear group introduced above: it is an open subset of the space of all n-by-n matrices, because it is given by the inequality

$\det(A) \neq 0$,

where A denotes an n-by-n matrix.

Lie groups are of fundamental importance in modern physics: Noether's theorem links continuous symmetries to conserved quantities. Rotation, as well as translations in space and time are basic symmetries of the laws of mechanics. They can, for instance, be used to construct simple models—imposing, say, axial symmetry on a situation will typically lead to significant simplification in the equations one needs to solve to provide a physical description.[v] Another example are the Lorentz transformations, which relate measurements of time and velocity of two observers in motion relative to each other. They can be deduced in a purely group-theoretical way, by expressing the transformations as a rotational symmetry of Minkowski space. The latter serves—in the absence of significant gravitation—as a model of space time in special relativity. The full symmetry group of Minkowski space, i.e. including translations, is known as the Poincaré group. By the above, it plays a pivotal role in special relativity and, by implication, for quantum field theories. Symmetries that vary with location are central to the modern description of physical interactions with the help of gauge theory.

GENERALIZATIONS

In abstract algebra, more general structures are defined by relaxing some of the axioms defining a group. For example, if the requirement that every element has an inverse is eliminated, the resulting algebraic structure is called a monoid. The natural numbers **N** (including 0) under addition form a monoid, as do the nonzero integers under multiplication (**Z** \ {0}, ·), see above. There is a general method to formally add inverses to elements to any (abelian) monoid, much the same way as (**Q** \ {0}, ·) is derived from (**Z** \ {0}, ·), known as the Grothendieck group. Groupoids are similar to groups except that the composition a • b need not be defined for all a and b. They arise in the study of more complicated forms of symmetry, often in topological and analytical structures, such as the fundamental groupoid or stacks. Finally, it is possible to generalize any of these concepts by replacing the binary operation with an arbitrary n-ary one (i.e. an operation taking n arguments). With the proper generalization of the group axioms this gives rise to an n-ary group. The table gives a list of several structures generalizing groups.

Group-like structures					
	Totality$^{\alpha}$	**Associativity**	**Identity**	**Invertibility**	**Commutativity**
Semicategory	Unneeded	Required	Unneeded	Unneeded	Unneeded
Category	Unneeded	Required	Required	Unneeded	Unneeded
Groupoid	Unneeded	Required	Required	Required	Unneeded
Magma	Required	Unneeded	Unneeded	Unneeded	Unneeded
Quasigroup	Required	Unneeded	Unneeded	Required	Unneeded
Loop	Required	Unneeded	Required	Required	Unneeded
Semigroup	Required	Required	Unneeded	Unneeded	Unneeded
Monoid	Required	Required	Required	Unneeded	Unneeded
Group	Required	Required	Required	Required	Unneeded
Abelian group	Required	Required	Required	Required	Required
^α Closure, which is used in many sources, is an equivalent axiom to totality, though defined differently.					

REFERENCES

1. Boyer, Carl B. (1991), A History of Mathematics (Second ed.), John Wiley & Sons, Inc., ISBN 0-471-54397-7
2. Donald R. Hill, Islamic Science and Engineering (Edinburgh University Press, 1994).
3. Ziauddin Sardar, Jerry Ravetz, and Borin Van Loon, Introducing Mathematics (Totem Books, 1999).
4. George Gheverghese Joseph, The Crest of the Peacock: Non-European Roots of Mathematics (Penguin Books, 2000).
5. John J O'Connor and Edmund F Robertson, History Topics: Algebra Index. In MacTutor History of Mathematics archive (University of St Andrews, 2005).
6. I.N. Herstein: Topics in Algebra. ISBN 0-471-02371-X
7. R.B.J.T. Allenby: Rings, Fields and Groups. ISBN 0-340-54440-6
8. L. Euler: Elements of Algebra, ISBN 978-1-899618-73-6
9. Asimov, Isaac (1961). Realm of Algebra. Houghton Mifflin.
10. Artin, Michael (1991), Algebra, Prentice Hall, ISBN 978-0-89871-510-1, Chapter 2 contains an undergraduate-level exposition of the notions covered in this article.
11. Devlin, Keith (2000), The Language of Mathematics: Making the Invisible Visible, Owl Books, ISBN 978-0-8050-7254-9, Chapter 5 provides a layman-accessible explanation of groups.
12. Fulton, William; Harris, Joe (1991). Representation theory. A first course. Graduate Texts in Mathematics, Readings in Mathematics. **129**. New York: Springer-Verlag. ISBN 978-0-387-97495-8. MR 1153249, ISBN 978-0-387-97527-6..
13. Hall, G. G. (1967), Applied group theory, American Elsevier Publishing Co., Inc., New York, MR 0219593, an elementary introduction.
14. Herstein, Israel Nathan (1996), Abstract algebra (3rd ed.), Upper Saddle River, NJ: Prentice Hall Inc., ISBN 978-0-13-374562-7, MR 1375019.
15. Herstein, Israel Nathan (1975), Topics in algebra (2nd ed.), Lexington, Mass.: Xerox College Publishing, MR 0356988.
16. Lang, Serge (2002), Algebra, Graduate Texts in Mathematics, **211** (Revised third ed.), New York: Springer-Verlag, ISBN 978-0-387-95385-4, MR 1878556
17. Lang, Serge (2005), Undergraduate Algebra (3rd ed.), Berlin, New York: Springer-Verlag, ISBN 978-0-387-22025-3.
18. Ledermann, Walter (1953), Introduction to the theory of finite groups, Oliver and Boyd, Edinburgh and London, MR 0054593.
19. Ledermann, Walter (1973), Introduction to group theory, New York: Barnes and Noble, OCLC 795613.
20. Robinson, Derek John Scott (1996), A course in the theory of groups, Berlin, New York: Springer-Verlag, ISBN 978-0-387-94461-6.
21. Artin, Michael (1991). Algebra. Prentice-Hall.

22. Atiyah, Michael; Macdonald, Ian G. (1969). Introduction to commutative algebra. Addison–Wesley.
23. Bourbaki, N. (1998). Algebra I, Chapters 1-3. Springer.
24. Cohn, Paul Moritz (2003), Basic algebra: groups, rings, and fields, Springer, ISBN 978-1-85233-587-8.
25. Eisenbud, David (1995). Commutative algebra with a view toward algebraic geometry. Springer.
26. Gardner, J.W.; Wiegandt, R. (2003). Radical Theory of Rings. Chapman & Hall/ CRC Pure and Applied Mathematics. ISBN 0824750330.
27. Herstein, I. N. (1994) [reprint of the 1968 original]. Noncommutative rings. Carus Mathematical Monographs. **15**. With an afterword by Lance W. Small. Mathematical Association of America. ISBN 0-88385-015-X.
28. Jacobson, Nathan (2009). Basic algebra. **1** (2nd ed.). Dover. ISBN 978-0-486-47189-1.
29. Jacobson, Nathan (1964). "Structure of rings". American Mathematical Society Colloquium Publications (Revised ed.). **37**.
30. Jacobson, Nathan (1943). "The Theory of Rings". American Mathematical Society Mathematical Surveys. **I**.
31. Kaplansky, Irving (1974), Commutative rings (Revised ed.), University of Chicago Press, ISBN 0-226-42454-5, MR 0345945.
32. Lam, Tsit Yuen (2001). A first course in noncommutative rings. Graduate Texts in Mathematics. **131** (2nd ed.). Springer. ISBN 0-387-95183-0.
33. Lam, Tsit Yuen (2003). Exercises in classical ring theory. Problem Books in Mathematics (2nd ed.). Springer. ISBN 0-387-00500-5.
34. Lam, Tsit Yuen (1999). Lectures on modules and rings. Graduate Texts in Mathematics. **189**. Springer. ISBN 0-387-98428-3.
35. Lang, Serge (2002), Algebra, Graduate Texts in Mathematics, **211** (Revised third ed.), New York: Springer-Verlag, ISBN 978-0-387-95385-4, Zbl 0984.00001, MR 1878556.
36. Matsumura, Hideyuki (1989). Commutative Ring Theory. Cambridge Studies in Advanced Mathematics (2nd ed.). Cambridge University Press. ISBN 978-0-521-36764-6.
37. Milne, J. "A primer of commutative algebra".
38. Rotman, Joseph (1998), Galois Theory (2nd ed.), Springer, ISBN 0-387-98541-7.
39. van der Waerden, Bartel Leendert (1930), Moderne Algebra. Teil I, Die Grundlehren der mathematischen Wissenschaften, **33**, Springer, ISBN 978-3-540-56799-8, MR 0009016MR 0037277MR 0069787MR 0122834MR 0177027MR 0263581.
40. Wilder, Raymond Louis (1965). Introduction to Foundations of Mathematics. Wiley.
41. Zariski, Oscar; Samuel, Pierre (1958). Commutative Algebra. **1**. Van Nostrand.

42. Definition 4.1.1, p181, Chapter 4.1 FIELDS: ROOTS OF POLYNOMIALS, Abstract Algebra, 3rd Edition (2006), by John A. Beachy, William D. Blair, Northern Illinois University, Waveland Press INC. www.waveland.com, ISBN 1-57766-443-4
43. Wallace, D A R (1998) Groups, Rings, and Fields, SUMS. Springer-Verlag: 151, Th. 2.
44. Second row at p120, Chapter 3 GROUPS, Abstract Algebra, 3rd Edition (2006), by John A. Beachy, William D. Blair, Northern Illinois University, Waveland Press INC. www.waveland.com, ISBN 1-57766-443-4
45. p120, Chapter 3 GROUPS, Abstract Algebra, 3rd Edition (2006), by John A. Beachy, William D. Blair, Northern Illinois University, Waveland Press INC. www.waveland.com, ISBN 1-57766-443-4
46. Karl Georg Christian v. Staudt, Beiträge zur Geometrie der Lage (Contributions to the Geometry of Position), volume 2 (Nürnberg, (Germany): Bauer and Raspe, 1857). See: "Summen von Würfen" (sums of throws), pp. 166-171 ; "Produkte aus Würfen" (products of throws), pp. 171-176 ; "Potenzen von Würfen" (powers of throws), pp. 176-182.
47. Peter Gustav Lejeune Dirichlet with R. Dedekind, Vorlesungen über Zahlentheorie von P. G. Lejeune Dirichlet (Lectures on Number Theory by P.G. Lejeune Dirichlet), 2nd ed., volume 1 (Braunschweig, Germany: Friedrich Vieweg und Sohn, 1871), p. 424. From page 424: "Unter einem Körper wollen wir jedes System von unendlich vielen reellen oder complexen Zahlen verstehen, welches in sich so abgeschlossen und vollständig ist, dass die Addition, Subtraction, Multiplication und Division von je zwei dieser Zahlen immer wieder eine Zahl desselben Systems hervorbringt." (By a "field" we will understand any system of infinitely many real or complex numbers, which is so closed and complete that the addition, subtraction, multiplication, and division of any two of these numbers always again produces a number of the same system.)
48. J J O'Connor and E F Robertson, The development of Ring Theory, September 2004.
49. Moore, E. Hastings (1893), "A doubly-infinite system of simple groups", Bulletin of the New York Mathematical Society, *3* (3): 73–78, doi:10.1090/S0002-9904-1893-00178-X, JFM 25.0198.01. From page 75: "Such a system of s marks [i.e., a finite field with s elements] we call a field of order s."
50. Earliest Known Uses of Some of the Words of Mathematics (F)
51. Fricke, Robert; Weber, Heinrich Martin (1924), Lehrbuch der Algebra, Vieweg, JFM 50.0042.03
52. Steinitz, Ernst (1910), "Algebraische Theorie der Körper", Journal für die reine und angewandte Mathematik, ***137***: 167–309, doi:10.1515/crll.1910.137.167, ISSN 0075-4102, JFM 41.0445.03
53. Jacobson (2009), p. 213
54. Jacobson (2009), p. 213

Index